ROUTLEDGE LIBRARY EDITIONS:
JAPAN

BROADCASTING IN JAPAN

BROADCASTING IN JAPAN

MASAMI ITO

With a Study Group on the Japanese Broadcasting System
HIROSHI SHIONO, TOSHIO KATAOKA AND
SEIGO NAGATAKE

Volume 8

Routledge
Taylor & Francis Group
LONDON AND NEW YORK

First published in 1978

This edition first published in 2011
by Routledge
2 Park Square, Milton Park, Abingdon, Oxon, OX14 4RN

Simultaneously published in the USA and Canada
by Routledge
711 Third Avenue, New York, NY 10017, USA

Routledge is an imprint of the Taylor & Francis Group, an informa business

British Library Cataloguing in Publication Data
A catalogue record for this book is available from the British Library

ISBN 13: 978-0-415-56498-4 (Set)
eISBN 13: 978-0-203-84317-8 (Set)
ISBN 13: 978-0-415-58518-7 (Volume 8)
eISBN 13: 978-0-203-84568-4 (Volume 8)

Publisher's Note
The publisher has gone to great lengths to ensure the quality of this reprint but points out that some imperfections in the original copies may be apparent.

Disclaimer
The publisher has made every effort to trace copyright holders and would welcome correspondence from those they have been unable to trace.

BROADCASTING IN JAPAN

MASAMI ITO
WITH A STUDY GROUP ON
THE JAPANESE BROADCASTING SYSTEM

CASE STUDIES ON BROADCASTING SYSTEMS

ROUTLEDGE & KEGAN PAUL
London, Henley and Boston
in association with the
INTERNATIONAL INSTITUTE OF COMMUNICATIONS

First published in 1978
by Routledge & Kegan Paul Ltd
39 Store Street,
London WC1E 7DD,
Broadway House,
Newtown Road,
Henley-on-Thames, Oxon RG9 1EN and
9 Park Street,
Boston, Mass. 02108, USA
Printed in Great Britain by
Redwood Burn Ltd.
Trowbridge and Esher

British Library Cataloguing in Publication Data

Ito, Masami
Broadcasting in Japan. - (Case studies on
broadcasting systems).
1. Broadcasting - Japan
I. Title II. Series
384.54'0952 HE8689.9.J/ 78-40551

ISBN 0 7100 0043 X

CONTENTS

FOREWORD

In many different parts of the world official and unofficial enquiries, often protracted, are being carried out concerning the future of broadcasting. Some of them have recently been completed. In every case two points have almost immediately become clear. First, the future of broadcasting can never be completely separated from its past, even though the history of broadcasting in all countries is a recent one: there may be sharp breaks, not all of which are registered in legislation, but there are also continuities. Second, the future, like the past, will not depend on technological development alone. There are many exciting new communications technologies, many of them still in their early stages, but the speed and scope of their development will be determined by social, economic, political and cultural factors as well as by the technologies themselves. It has always been so.

Common technologies have been employed in different ways in different countries - sometimes with a few measures of control imposed by governments, by professional groups, or by trade unions, often with many. It is remarkable to what a great extent it is necessary to understand the general history of particular countries in order to understand what they have done with their conscious or unconscious communications policies.

This series of monographs, sponsored by the International Institute of Communications, is intended to direct attention to the main features of the communications patterns of a number of different countries. The studies deal with broadcasting structures rather than with the detailed processes of programme making or with the diffusion of news and ideas; and they seek first to explain how these structures came into existence, second, to identify what have been the landmarks in their histories, and, third, to elucidate what are the alternative possibilities envisaged for the future. Of course, a knowledge of the structures by itself is not enough to enable an adequate evaluation to be made of the quality of broadcasting output in any particular case. The same structure will generate different output at different times, and very similar structures will generate very different outputs.

Until recently it was thought possible to distinguish broadly between on the one hand systems controlled by government and on the other hand systems linked with business through private enterprise and advertising. Yet there was always a third type of system, represented formidably by the BBC, which entailed neither government control nor business underpinning. This system, which was widely copied, was seldom copied in its entirety, and it now has many variants, most of which have deviated substantially from the model. In many countries there are now dual or multiple systems, in some cases, but not in all, subject to common 'supervision'- and in all countries there are degrees and nuances of control of broadcasting output whether by governments or by market forces.

The United States system, which is important not only in itself but because of the influence it has through exports of programmes and through diffusion of broadcasting styles, is itself a complex system - containing as it does a multiplicity of agencies and a changing public service element. It is hoped that United States experience will be covered in a later volume. There is increasing pressure there for a major review in the light of that experience and of continuing technological change affecting not only broadcasting but a wide and increasingly interrelated group of new communications technologies.

Alongside complex national structures, the products of time and place and in many cases deeply resistant to fundamental change, there are, of course, many new broadcasting structures in the world, including many which have come into existence in recent years in new countries. Many of these structures reveal themselves as extremely complex, too, when they are subjected to careful scrutiny. Nor are they necessarily very malleable. The more governments set out to chart and carry through conscious 'communications policies' - often related directly to their planning policies - the more they are compelled to consider the relationship of 'traditional' modes of communication to new technologies. The more, too, they are forced to establish priorities. This series includes, therefore, a number of countries where such policies have been formulated or are in the course of formulation.

Measuring the distance between policy formulation and policy implementation or effectiveness is, of course, at least as difficult in this field as in any other, and interesting work is being carried out by scholars in several countries on promise and performance. This series of studies, however, is less ambitious in intention. The studies are designed to provide accessible and reliable information rather than to evaluate the quality of achievement. The first cases chosen include some where there is no existing manageable monograph and some where the particular experience of that country is of general interest at the present time. The countries selected include some which are old and some which are new, some which are big and some which are small.

As the series unfolds, there will be increasing scope for comparison and contrast, and international patterns will doubtless be revealed - of 'models', of 'imports' and 'exports', of regional 'exchanges', and of relationships between different media. As such comparison and contrast become more sophisticated than they have been in the past, any conclusions reached will be of increasing value in the future to those policy makers who are concerned to see their own circumstances in perspective and to frame their choices clearly.

Meanwhile, the International Institute of Communications, formerly called the International Broadcast Institute, which first launched this series, will continue to concern itself with the general opportunities and problems associated with the continuing advance of communications technology. The Institute is an international body which seeks to bring together engineers and social scientists, lawyers and programme-makers, academics and administrators.

The author of each case study in this series has been free to assemble and to present material relating to his own country in a manner decided upon by him, and he alone is responsible for the evidence offered and for the conclusions reached. Yet guidelines have been given him about arrangement and coverage. Thus, he has been encouraged to ask questions as well as to compile facts. What have been the critical points in the history of broadcasting? How have that history and the broadcasting structures which have been evolved been related to the history of other forms of communication (the Press, for example)? What are the main institutional relationships at the present time? What are likely to be the future trends? Is it possible to talk of an integrated 'communications policy' in the case of the country under review?

The International Institute of Communications has no views of its own as an institution on the answers to such questions, but its Trustees and members believe that answers should be forthcoming if debate is to be both lively and well-informed. Much of the serious study of communications systems has hitherto been carried out within the confines, cultural as well as political, of national boundaries, and it is such research which most easily secures financial support. This series will point in a different direction. It is not only comparison and contrast which are necessary but a grasp of what problems and opportunities are common to countries, not necessarily alone but in the great continental broadcasting unions or other groupings between states.

We can now trace the beginnings of a 'global' sense in communications studies. Indeed, the word 'beginning' may be misleading. The sense certainly long preceded the use of satellites and was anticipated in much of the nineteenth-century literature. The world was being pulled together; it was becoming a smaller place; everyone, everywhere, it was suggested, would be drawn in, instantaneously.

Communications policies, of course, have often failed to unite: instead, they have pulled people apart in clashes of images as well as in wars of words. And some of the case studies in this series will show how.

Two final points should be made. First, nothing stands still in communications history, and there are bound to be changes between the writing of these case studies and their publication. The processes of implementation of policy changes are often protracted. Second, it may well be that we are moving out of the age of 'mass broadcasting' as we have understood it into a new age of electronic communication. In that case, these studies will appear at a strategic time and will deserve careful study separately and together.

Asa Briggs
Worcester College, Oxford

ACKNOWLEDGMENTS

The International Institute of Communications is most grateful to the Hōsō-Bunka Foundation, Inc. for its assistance in financing this series of case studies on broadcasting systems.

Chapter 1

ENVIRONMENT FOR BROADCASTING

Since its emergence as a modern state after the Meiji Restoration, Japan has steadily marched toward the development of its own culture and the construction of a modern society, merging the newly introduced Western culture and the traditional Eastern culture already deeply rooted in Japanese soil. This march was temporarily checked by Japan's defeat in World War II. Since that time, however, Japan has achieved rapid economic growth and has become the only Oriental country which is highly industrialized and urbanized. In its high level of technology and education, and in its relatively stable political structure, Japan can now be favourably compared with advanced European countries and America. Japan today may well be said to belong to the Far West as well as to the Far East.

1 BASIC FEATURES

The Japanese archipelago consists of more than 1,000 inhabitable islands including the four main islands of Honshu, Hokkaido, Kyushu and Shikoku. With a total area of 372,000 square kilometres and a population exceeding 104 million, the population density is approximately 281 persons per square kilometre. Moreover, the major part of the country is mountainous, and land available for cultivation makes up only 15 per cent of the total. The population density per arable land, therefore, is the highest in the world: 1,580 people per square kilometre. With the growth of industrial activity, more and more people have come to live in urban areas, creating a sharp distinction between densely and sparsely populated regions. In particular, the megalopolitan Tokaido area, the adjacent Kanto district facing the Pacific, and the Kinki area to the west, have an extremely high concentration of inhabitants and industries. It may well be said, therefore, that the Japanese people live in overcrowded conditions, in a limited area within a small expanse of land.

There was a marked increase in the birth rate soon after the war's end, and the peak in the population curve by age groups shows that young people in their twenties are now the largest number. Quite recently, the birth rate has shown an increase of 18.8 per million, but generally, the ratio of children under 14 has decreased since the war. At present, it should be noted that the ratio of people of productive age is high. On the other hand, with the very low death rate, the ratio of old people is predicted to increase rapidly in the near future.

Ethnically, the Japanese belong to the Mongolian race, and despite various local distinctions, they are considered to belong to a single race exhibiting almost uniform features. The Japanese language is rich in dialects, but obviously has its origin in a single parent language. This has led to a homogeneity in culture, and also partly accounts for the speedy growth of mass media in modern Japanese society.

Pre-war Japanese society was characterized by the coexistence of both the modern sector, developed in big cities, and the traditional sector which survived in the agricultural areas. At present, however, Japanese society is highly industrialized and urbanized and its traditional sector has almost disappeared, even in the rural communities. It follows from this that the social, economic and cultural life of the Japanese people has been standardized and the gap between the rich and the poor is not so great. This social pattern also contributes to the development of the mass media.

The Japanese islands stretch in a semi-circular arc extending north to south, the northern tip lying at latitude 45° 31' N. and the southern end at latitude 34° 3' N. The land varies in the height above sea level, the sea coasts are washed by both warm and cold currents, and the mountain range running lengthwise down the country divides it roughtly into two climatic zones, one facing the Pacific and the other the Japan Sea. Hence there is a sharp climatic difference from region to region. The weather is also subject to abrupt and complex changes caused by frequent typhoons and travelling masses of high air pressure. Natural disasters resulting from typhoons, and heavy rains in the wet season are not infrequent. On the whole, however, the climate is mild and comfortable, with regular seasonal changes.

2 OUTLINE OF POLITICS AND THE ECONOMY

Japan is a parliamentary democracy with sovereign power resting with the people. The Imperial system, however, grounded as it is in history and traditions still exists. In former times the Emperor held complete sovereignty, but under the present Constitution, serves now only as a symbol uniting the state and the people. He has no power in government, and performs only formal and ceremonial functions.

The bicameral Diet, the supreme body exercising national power, is composed of the House of Representatives (511 in all) and the House of Councillors (252 in all). Members of both Houses are elected directly by the people, although there is a distinction between the two Houses regarding constituencies, terms of office, eligible age for candidacy, dissolution, etc. The House of Representatives has greater power than the House of Councillors. For instance, the House of Representatives may overthrow the Cabinet by passing a vote of no confidence. When the decisions of the Houses conflict, the House of Representatives may enact a bill by passing it again with a two-thirds majority.

The Japanese Cabinet is answerable to the Diet, as under the British Parliamentary Cabinet system, and the Prime Minister is nominated from among the Diet members. In practice, the Cabinet is formed by the political party which occupies the majority of the Lower House seats. The Cabinet is composed of not more than 19 Ministers of State including the Prime Minister. The Constitution provides that the majority of the Cabinet members should be Diet members, and with very few exceptions, all Cabinet members are Diet members. Under the supervision of the Cabinet, which is a consultative body, administration of the country is carried out by Ministries, headed by Ministers of State, and by government Agencies. Government officials who work under the Ministers are non-political in status. Because of their excellent abilities they have proved themselves to be the main pillar of Japan's administration.

Judicial power is exercised by courts of justice headed by the Supreme Court. Unlike France or Germany, Japan has no specific court for administrative litigation. The courts, in addition to ordinary business, are empowered to judge constitutional cases, and thereby serve as a watchdog to see that both the legislative and administrative organs act in conformity with the Constitution.

The most important of their functions is to safeguard human rights, including freedom of expression and speech for the press, broadcasting and other mass media, guaranteed under the Japanese Constitution. This forms the basis of Japanese law concerning mass media.

In addition to the central government, there is also a system of local autonomy, administered by local governments and municipal authorities, etc. The heads of these organs, as well as local assembly men, are elected by popular vote. Various systems of direct democracy have also been established.

Political parties work as the prime movers of these political structures. In 1955, the merger of various political parties brought about a duopoly of the Liberal Democratic Party and the Socialist Party. Later, both progressive and conservative parties increased in number. At present, the Liberal Democrats, the Socialists, the Komei Party, the Democratic Socialist Party and the New Liberal Club hold seats in the Lower House. Immediately after the war the premiership was held by a Socialist for a short time. With this single exception, however, the Liberal Democrats (or their precursors) have held power and, in this sense, the political situation has been stable. Recently, however, the Liberal Democrats have been finding it increasingly difficult to maintain their one-party majority in both Houses. A popular movement against the conservative administration, characterized by its close links with bureaucracy and business interests, seems to be gaining momentum.

The Japanese economy, which collapsed following Japan's surrender in World War II, has made a surprising recovery in the past thirty years. In 1958, the Gross National Product stood at no more than $23.5 billion, but by 1971, it had risen to $208.4 billion, making Japan a major economic power ranking next to the USA and the Soviet Union. The per capita annual income, however, indicative of the people's standard of living, is, at $3,770, still below the level of other advanced nations.

The rapid economic growth witnessed in Japan is ascribed to various factors: concentrated investment in production facilities; a strong propensity for saving among the people; the existence of an extremely diligent and highly educated work force; the positive introduction of new technology from abroad; appropriate collaboration between bureaucracy and private industries; workers' resolute loyalty to their companies, working under a system of lifelong employment and advancement by seniority that ensures their financial security; close co-operation between banks and private companies which, through the resultant easing of bank loans, has contributed greatly to industrial development; the comparative stability of politics, and the peace-oriented Japanese Constitution under which non-productive defence expenditures have been kept very low in comparison with other advanced nations.

Another factor is the change in Japan's industrial structure in response to international economic trends. In prewar days, the major industries of Japan were light industries, centring around spinning. In postwar years, however, the emphasis has shifted to heavy, chemical industries, and metals, machinery and chemical goods now represent the major industrial fields. In 1972, 79.3 per cent of exports were from the heavy industrial section. It may safely be said that Japan has been much quicker in its response to shifting world demands than have other countries. In the area of broadcasting, the progress in electronic industries has been remarkable. Foreign technology has been rapidly introduced and assimilated. The sale of new products such as television sets and transistor radios first marketed in Japan later greatly expanded on the foreign markets. These products now constitute one of Japan's principal industries. As a result, however, the latter half of the 1960s saw the surfacing of environmental pollution problems. The oil shock of 1973 and its after-effects have deprived Japan of its past dream of rapid economic growth. How to deal with these problems is a serious task confronting the future Japanese economy.

3 OUTLINE OF THE EDUCATIONAL SYSTEM

Japan suffers from an acute shortage of natural resources, and thus education for the development of human abilities has long been accorded great importance. Prewar education was based on traditional Japanese educational thought. After World War II, however, educational reforms were introduced under the guidance of the Occupation Forces and led to the establishment of the present educational system controlled by the Ministry of Education. The prewar educational administration was strongly coloured by its emphasis on supervision, but today the Ministry of Education is concerned more with the promotion and diffusion of education, and restricts its supervisory activities to guidance and advice, and setting national educational standards. To prevent centralization each local government has its own Board of Education charged with the administration of local education. Initially, members of the Board were elected by popular vote, but currently they are appointed by the heads of local public bodies with the consent of local assemblies.

The system consists basically of six years of primary school, three years of junior high school, three years of senior high school, and four years of university or college. Elementary education at the primary and junior high school level is compulsory and free with the exception of private schools. Public schools constitute the overwhelming majority. Textbooks used are authorized by the Ministry of Education and supplied to students free of charge. The rate of school attendance is 99 per cent plus, the few non-attenders have special exemptions.

Senior high schools, linked with junior high schools, provide advanced secondary education. In addition to day schools, there are also night or correspondence schools. Almost all pupils attend senior high school, particularly in urban areas.

Universities, prior to the inauguration of the new school system, were comparatively few in number (18 state universities, 3 universities established by local governments, and 27 private universities), but today there are 423 universities and colleges (83 state universities, 33 universities established by local governments and 307 private ones). University enrolment has rapidly increased. Including junior colleges, the figure comes to nearly 40 per cent. It is becoming increasingly difficult to cope with the surplus of applicants merely by establishing more universities or expanding existing school buildings. In this connection, the idea of establishing an 'air' university, using broadcasting as the principal means of instruction, is being advanced.

In addition to this main stream of school education, there are kindergartens for pre-school children over three years of age, junior colleges with two-year courses designed to provide practical and occupational knowledge and training (of the total of 511 junior colleges, 411 are private, with mainly female students), and higher professional schools for the training of technicians (graduates from junior high eligible for a five year course), as well as post-graduate schools for training advanced students and providing higher professional education (the Master's course being two years and the Doctor's course normally five years).

In addition, great importance is attached to offering a wide range of social adult education, primarily for working youths and housewives, conducted at local culture centres, youth hostels, etc.

4 OUTLINE OF TELECOMMUNICATIONS

After World War II, reconstruction of destroyed telecommunications facilities proceeded apace, initially under the unified control of the government Communications Ministry. In 1949, however, the Ministry was divided into the Ministry of Postal Services and the Ministry of Telecommunications. In 1952, an epochal reorganization brought into being the Nippon Telegraph and Telephone Public Corporation which, by replacing the Ministry of Telecommunications, launched the monopolistic and unified management and operation of all domestic telecommunications. The new public corporation under the supervision of the re-established Ministry of Posts and Telecommunications has the responsibility for the provision of all services and for their efficient operation on a cost recovery basis.

Based on long-range plans, the Telephone and Telegraph Corporation has made great efforts in advancing telecommunications, particularly telephones. The number of telephones installed by the end of 1974 was 41,900,000, second only to the USA. In spite of the outstanding progress of the past ten years, however, the number of telephones per household is, as yet, only 62.9 per 100, still a smaller rate than Europe or America. Expansion has been slow particularly in farming and fishing areas. However wire broadcasting and telephone services are widely utilized in some areas, and meet what is a pressing demand. At present various measures are being taken to improve the welfare of solitary, aged and physically handicapped persons by extending the telephone service and developing special telephone devices.

The use of the telegraph system, once considered an important means of communication has been declining with the advance of the telephone. From a peak around 1960, the use of telegrams has gradually declined, an increasing proportion being used for congratulations and condolences.

In addition to these public service telecommunications, rapid progress recently made in the field of radio science has expanded the use of telecommunications in various forms. Privately-owned telecommunications devices, connecting at fixed points, are being used on an increasingly wide scale - a labour-saving and rationalization measure for industry - and play an important role through the national or local network of microwave or shortwave circuits. Greater use is also being made of mobile air, maritime and land communications for purposes of traffic control and promotion of industrial activity. Data conveying communications over computerized telecommunication circuits have made notable advances in the past several years. In this connection, it is currently being urged that medical information, daily consumers' information and other service systems be developed on a nationwide scale, enhancing the well-being of the individual and society as a whole, at the same time paying due attention to the protection of privacy.

Apart from domestic telecommunications the Kokusai Denshin Denwa Co, Ltd was established in 1953 to operate international telecommunications on a monopolistic basis. The company is now actively involved in satellite communication.•

5 OUTLINE OF MASS MEDIA OTHER THAN BROADCASTING

Mass media reached maturity in this country in the 1920s. This rapid growth was possible because the populace enjoys a high standard of education in an ethnically and linguistically homogeneous society. In prewar days there was a wide circulation of newspapers and periodicals, and radio broadcasts were enjoyed by large audiences. In the postwar Occupation era, the removal of prewar restrictions on freedom of speech and the press, and the abolition of censorship, accorded to mass media a freedom safeguarded by law. Thereafter, rapid progress was made, sustained by a society in which the voice of the masses is strong. Present-day Japan can be said to be a nation with a giant mass media system, occupying a leading position among nations in the field of mass communications.

Japanese newspapers can be classified into three types: national, regional newspapers covering more than one prefecture, and local, with distribution confined within a single prefecture. There is a total circulation of about 60.78 million, with 1.85 readers per copy, and 1.79 copies per household. This circulation rate is the second highest in the world, next only to Sweden. The five national papers account for more than half the total, and their editorials are very influential. In comparison to their counterparts abroad they are financially stable. No financial aid is accepted from the government to exclude any encroachment on the freedom of the press.

Distinctive characteristics of Japanese newspapers include the subscription fee, which covers the cost of both morning and evening editions for large cities and their vicinities; a home delivery service, operated by 98 per cent of regular newspapers; and their non-partisan policy, free of any political colouring or ideology.

Most importantly, newspapers occupy a central position in mass media as a whole, and constitute a comprehensive information industry. The national newspapers also publish monthly and weekly magazines, operate publishing businesses, have investments or hold shares in commercial broadcasting companies, sponsor sports meetings and undertake such cultural enterprises as public exhibitions. Not a few local newspapers also have close relationships with local commercial broadcasters in some form or other. Newspapers draw on the services of the two Japanese news agencies, the Kyodo News agency and the Jiji News agency.

Weeklies, initially the monopoly of newspaper companies, were published for the first time in 1956. Their special features depicting social customs and manners soon proved popular and they are now an established mass media form. They range from quality magazines to cheap comic magazines, and down to those appealing to more vulgar tastes. As of 1975, there were 56 weeklies in circulation, reaching a total of about 1,176 million copies. In general, they provide light entertainment and newsy material to much of society.

Monthly magazines total 1,446, with a circulation of about 1,238 million. These include children's magazines (27.3 per cent) and also practical knowledge magazines (10.2 per cent). Women's magazines, which once enjoyed a mammoth circulation, now tend to be on the decrease. Similarly, the so-called 'general' magazines which once played a forceful role in the formation of public opinion, are also on the decline.

Books are published in large numbers as well, the total number of issues to date amounting to 733.2 million. According to UNESCO figures, Japan ranks fifth in the world in book publication.

In terms of the number of films produced domestically and the number of movie-goers, motion pictures in Japan reached their peak in 1958. Some Japanese films of this era have won international acclaim. Since then, however, the film industry has declined, and in spite of financial support from the government and a reduction in entertainment taxes, there has been no visible sign of recovery. Two primary reasons for this decline can be cited; first, the increased popularity of television, bringing many films to the home television screen and resulting in a decrease in the number of movie-goers; and second, an increased variety and volume of entertainment, which has deprived the film world of its former glory acquired in days of scarce entertainment.

Outlay for advertising in Japan, in the 1975 fiscal year, amounted to approximately ¥1,237.5 billion. The advertising industry has made progress in close relation with mass media. Once it was the newspaper that constituted the leading medium for advertising, but, as of 1975, advertising through television rose to 34 per cent of the total, overtaking the ratio of the newspaper which stood at 33.1 per cent. This is but one further indication of how far television has advanced in the field of mass media.

Chapter 2

THE EVOLUTION OF BROADCASTING IN JAPAN

1 FROM THE INCEPTION OF BROADCASTING TO THE END OF WORLD WAR II

(a) Inception of radio broadcasting

Radio broadcasting in Japan was begun by the Tokyo Broadcasting Station, a public utility corporation, on 22 March 1925. The only law governing the use of radio waves at that time was the 1915 Wireless Telegraphy Law. This law provided that the use of radio frequencies should be regulated by government in accordance with the necessary requirements of different services. According to the law, authorization could be given for private operation of radio communication stations.

For that reason, in June 1922, the Ministry of Communications began to look into the question of the supervision and regulation of broadcasting services, taking into account the pressures from the private sector for their operation. The Great Kanto Earthquake of 1 September 1923 with its devastation and confusion drove home to government authorities the importance of broadcasting immediate and relevant information to the public. The Ministry of Communications therefore took prompt steps to establish an administrative system for radio broadcasting. In the same year, the Ministry instituted an ordinance concerning the authorization and supervision of radio broadcasting services. This, along with the 1915 Wireless Telegraphy Law, remained the only definitive broadcasting regulations until the year 1950.

The new regulations stimulated the launching of broadcasting services by private enterprise. To cope with this situation, the Ministry decided to entrust the management and operation of broadcasting services to a 'corporate juridical body for public service'. In 1924 the Tokyo Broadcasting Station, a public utility corporation, which was to cover the metropolitan area, and similar corporations were established in Nagoya and Osaka and each began broadcasting in 1925. The broadcasters were expected to finance their operations with receivers' fees, to be collected on the basis of contracts with the listeners. The collection of receivers' fees was to be endorsed by receivers' contracts issued by the government. Any listener who wished to apply for such a licence was required to submit to the government the written consent of the competent broadcaster (from 1926 a copy of a receivers' contract with competent broadcasters).

Broadcasters were not permitted to broadcast any matter banned from publication or the press under the laws then in force. All programmes were required to be submitted to the Ministry of Communications for advance censorship.

By the end of March 1925, 5,455 receivers' contracts had been confirmed in the metropolitan area alone. One year later, the figure had climbed to 258,507 (168,680 in the metropolitan area).

(b) Establishment of Nippon Hoso Kyokai (NHK)

Not long after the three public corporations began broadcasting operations, there arose demands for the expansion of the service to areas outside those covered by the existing stations.

In November 1925 the Tokyo Broadcasting Station submitted a proposal to the Ministry of Communications for increased output and for the inauguration of relay stations in all main regions except the Osaka and Nagoya areas. The Ministry conducted independent research and in February 1926 decided that nationwide radio broadcasting should be operated by a single corporate juridical body, merging with existing three.

As a result, the Nippon Hoso Kyokai (NHK) was established the same year, with the former three bodies working as branches of the new organization (later becoming central broadcasting stations).

The Articles of Incorporation as approved by the Ministry of Communications stipulated that by-laws relating to initial operations, those concerning operations in general, and such matters as the appointment and dismissal of directors and auditors, the inauguration and retirement of the president and managing directors, be subject to the approval of the Minister of Communications. The Corporation was thus placed under the strict supervision of the Ministry of Communications. In addition, it was ruled that the first officers of the Corporation be designated by the Minister of Communications. In approving the Articles of Incorporation, the Minister of Communications outlined the objectives of the programme for establishing radio broadcasting facilities and pointed out in his written directive that the technical construction programme, the year's business operation programme, the budgetary plan and numerous other matters should be submitted for approval or reported to the Minister.

(c) Extension of radio broadcasting and the initiation of the International Broadcasting Service

The completion of Nippon Hoso Kyokai (NHK)'s national radio network was pushed forward so as to meet the goal of providing nationwide relay of the coronation of the new Emperor Hirohito in November 1928. The number of receivers' contracts that year exceeded 500,000.

In 1930 NHK began the construction of local radio stations in major cities throughout the country. In 1930, the Corporation established its Technical Research Laboratories, and thereby launched full-scale studies of electronic technology, including television. The London Disarmament Conference in 1930 provided the occasion for the first successful international broadcast programme exchange among Japan, Britain and the USA.

NHK set up a second broadcasting network in Tokyo in 1931 and began regular education broadcasting. The stations in Osaka and Nagoya followed Tokyo's example in 1933 and this tripartite system continued through to 1941. Owing to a difference in views between the Ministry of Communications - in charge of broadcasting - and the Ministry of Education - charged with educational administration - however, it was not until 1935 that educational broadcasting was given due recognition in the school educational system.

In 1931, with the Manchurian Incident, the acquisition of radio sets proceeded apace and, in February 1932, the number of registered listeners reached the one million mark. From about this time government control over NHK's programming began to be tightened.

On 1 June 1935 NHK commenced its regular overseas service: 'Radio Tokyo'. Broadcasting was initially carried out for one hour each day in English and Japanese, and was directed towards the Western part of the USA. Thereafter, the overseas service was expanded both in terms of directions of reception and broadcasting hours. In 1936 an abortive attempt at a coup d'état by a section of young Army officers on 26 February (the 2.26 Incident) paved the way to Japanese militarism, and in this situation, the overseas service grew in scope.

On the day of this revolt, domestic radio reporting was banned by government order and NHK suspended most of its other broadcasts as well. Three days later, an NHK announcer read a radio message to the rebel soldiers, appealing for their surrender. That this broadcast contributed to the solution of the incident marked an epoch in the history of radio broadcasting.

In July 1936 an information committee was set up within the Cabinet, acting as a unified information and publicity liaison centre for the various Government Ministries.

On 7 July 1937 a clash between Japanese and Chinese forces escalated into full scale hostilities. Consequently, radio broadcasting played an increasingly important role in the publicizing of national policy. In September of the same year, the Cabinet Information Committee was reorganized into a Cabinet Information 'Division', taking charge of intelligence, internal and external information, and instruction-propaganda on government policies.

In 1940 this Division was upgraded to the status of a 'bureau', with power to control broadcasting affairs being part of its overall duties to guide and control various fields of mass media. NHK was thus placed under the dual control of the Prime Minister and the Minister of Communications.

The Pacific War broke out on 8 December 1941. With the aim of bolstering air defence, the government immediately set up a Radio-wave Control Headquarters composed of representatives of the Ministry of Communications, the Army and Navy Ministries and the Information Bureau, and a previously prepared 'Essentials of Emergency Structure for Domestic Broadcasting' was put into effect. NHK not only beamed flash news on the opening of hostilities, but also took measures to limit its domestic broadcasting in principle to national network broadcasts originating from Tokyo, to suspend secondary broadcasts, and to discontinue weather reporting and so forth. On the following day, 9 December, the Government issues a radiowave control order by which the same frequencies were to be used throughout the country, reducing the output power for the Central Tokyo Broadcasting Station and controlling broadcast operations. Radio broadcasting thus became no more than an instrument for the raising of national fighting morale.

As air raids over Japan proper intensified, the production and supply of receivers fell markedly although receivers provided the only source of information during the air attacks. By the end of March 1945 receiving contracts numbered 7.5 million.

On 14 August 1945 the Government decided to accept the terms of the Potsdam Declaration of the Allied Powers calling for unconditional surrender. At noon, the following day, the Imperial Proclamation announcing the termination of the war was broadcast in pre-recorded form. It was the first ever broadcast by His Majesty the Emperor.

With the end of the Pacific War on 1 September came the end of the radiowave control, and, capacities permitting, broadcasting stations began operating again with the same frequencies and output power as they had in prewar years. On the same day, secondary broadcasts were also resumed,

2 BROADCASTING AFTER WORLD WAR II

(a) Broadcasting during the Allied Occupation

On 2 September 1945 with the signing of the instrument of surrender, Japan's sovereign power was placed under the control of the Supreme Commander for the Allied Powers (SCAP) and the General Headquarters of the SCAP (GHQ). Directive No. 1 (2 September) and Directive No. 2 (3 September) ordered the Japanese Government to preserve and operate telecommunication facilities in their existing condition. Broadcasting in foreign languages was prohibited and on 10 September NHK discontinued its overseas broadcasts.

On the same day GHQ issued the Japanese government with the memorandum 'Freedom of Speech and Press'. Its purpose was to reflect the stipulation of the Potsdam Declaration that the Japanese Government should remove all obstacles to the revival and strengthening of democratic tendencies among the Japanese people. In parallel with its emphasis on the respect of freedom of expression, GHQ made clear its intention to exercise certain controls considered necessary for the accomplishment of occupation objectives. In broadcasting GHQ took concrete steps toward radio control by issuing a 'Radio Code for Japan' on 22 September.

GHQ not only imposed censorship over NHK programming, but also gave guidance and advice in the planning and production of its programmes. Continuous nation-wide programming, by which NHK was to be on the air at all times, quiz programmes with audience participation, broadcast 'forums', etc., were subsequently initiated.

Within ten days of the receipt of the 10 September SCAP memorandum, the Communications Board (formerly the Ministry of Communications) formulated a plan for establishing new broadcasting organizations operated by private industries, in addition to NHK. On 25 September the Cabinet approved this proposition. This Government move gave rise to various initiatives for the establishment of commercial stations in major cities, which, however, failed to bear immediate fruit.

On 11 December GHQ issued a memorandum on the reorganization of the Japan Broadcasting Corporation (NHK) and ordered the Communications Board to form a citizens' advisory committee for NHK, assigned to select candidates for the Corporation presidency. GHQ also deprived the Board of Information of its authority over the management of NHK. By May 1946, after conducting a reshuffle of its officers, including the President, and with the revision of the Articles of Incorporation, the postwar organization of NHK was complete.

By March 1946 receivers' contracts stood at 5.7 million, and audiences were already showing a steady expansion.

On 15 June 1946 NHK resumed its research into television, and, on the same day, the Radio Culture Research Institute was established. GHQ's control over broadcasting gradually began to slacken.

(b) Start of the new broadcasting system

On 11 November 1946 the new Government was promulgated, giving clear statement to the principles of pacifism, the doctrine of popular sovereignty and respect for fundamental human rights. Article 21 guaranteed freedom of 'speech, press and all other forms of expression', and censorship was abolished. The Constitution was put into effect on 3 May 1947.

In the month before the new Constitution was promulgated GHQ instructed the Communications Ministry (re-established in July 1946) to revise existing laws and regulations regarding broadcasting and communications so as to reflect more the spirit of the new Constitution. On 1 November the Communications Ministry established a special council to work on amending the laws concerned. Close contact with GHQ was maintained while doing this work. On 18 June 1948, when the proposed law was submitted to the National Diet, the suggestions from the various divisions of GHQ were found to be inconsistent and the bill was withdrawn. Finally, on 22 December 1949 three related bills (a proposed Radio Law, Broadcast Law and Radio Regulatory Commission Establishment Law) were submitted to the Diet. The bill for the establishment of a Radio Regulatory Commission, which conformed with GHQ instructions as embodied in General MacArthur's letter, passed the Diet without amendment on 24 April 1950. The two other bills were enacted with some amendments on 26 April 1950. The three laws were promulgated on 2 May of the same year, to take effect from 1 June. Thereafter, 1 June every year is commemorated by broadcasters as Radio Day.

The Radio Law which replaced the old Wireless Telegraphy Law as the fundamental law regulating radio communications, aims at exercising control only in so far as it is necessary to ensure efficient and appropriate utilization of radio waves, based on the people's freedom of that utilization. This law also removed the necessity for listeners to obtain government permission.

The Broadcast Law initiated a dual system of broadcasting, with NHK as a strictly public-service organization on the one hand and the free, private broadcasting organizations on the other. Provision was made for the reorganization of NHK into a new corporate body in charge of public broadcasting services, with any necessary supervision concerning the operation of its services, finance and other matters, being maintained by the National Diet, representing the people.

The Radio Regulatory Commission Establishment Law was modelled on the independent regulatory commission system in the USA. Under its stipulations, the Commission was empowered to regulate the emission of radiowaves and supervise broadcasting, and was invested with quasi-judicial and quasi-legislative functions.

On 1 June 1950 when the three laws came into effect, the Nippon Hoso Kyokai Corporate Juridical Body was dissolved, and its rights and duties were taken over by the new Nippon Hoso Kyokai (NHK) of today, established under the new law. The abbreviation NHK has been in use since 1946 when the former corporation still existed. Under the provisions of the Broadcast Law, NHK was to be financed by receivers' fees collected under contracts concluded with individual listeners equipped with receivers capable of receiving standard broadcasts.

On 21 April 1951 the Radio Regulatory Commission granted provisional licences to sixteen commercial stations for the first time in Japan, two each in Tokyo and Osaka and each of the twelve others in other major cities. On 1 September 1951 the Chubu Nippon Broadcasting Company (CBC) in Nagoya and the Shin Nippon Broadcasting Company (NJB) in Osaka began broadcasting as the first commercial stations in Japan. Earlier, on 20 July, a National Association of Commercial Broadcasters, composed of recipients of provisional licences, was formed. In 1951 four more stations and, in 1952, nine other stations earlier granted provisional licences, went into operation. (One station had its licence cancelled.) Programming included newscasts made possible by the backing of the newspaper companies who also had provided financial support for the stations' inauguration, as well as other entirely new programmes of their own making. Coupled with commercials, their programming was soon favourable received by the public, and, thereafter, there was a gradual increase in the number of commercial stations.

In September 1951 the San Francisco peace treaty was signed, and postwar Japan embarked on the preparatory stage for re-entry into international society. On 1 February 1952 NHK resumed its overseas broadcasting service which, under the Broadcast Law, was known as an 'international broadcasting' service.

(c) Start and spread of television

Toward the end of 1948, soon after NHK had resumed its television research, several plans for the establishment of television organizations were drawn up, but did not bear fruit. On 26 May 1951 the House of Representatives adopted a resolution calling for an early start to television broadcasting. On 2 October 1951 the Nippon Television Network (NTV) filed an application for instituting television operations.

On the 27th of the same month NHK followed suit, making an application for television broadcasting in Tokyo, Osaka, Nagoya and other cities. In the meantime, the Radio Regulatory Commission made its decision on the standard frequency system for Japanese television. There were heated debates among various interests as to whether a six megahertz or a seven megahertz band should be adopted, and eventually the six megahertz band (the American system) was decided on. In the same year, a number of commercial radio broadcasters submitted applications for television operation.

In NHK's case, a bill for partial amendment of the Broadcast Law as submitted by members of the House of Representatives passed the Diet on 11 June 1952 and went into force on 17 June. This amendment enables NHK to collect receivers' fees for television as well as radio. The Radio Regulatory Commission announced its decision on 'Policy and Measures for licensing television broadcasters' on 31 July 1952 granting Japan's first provisional television to NTV (Tokyo). Action on applications made by NHK (Tokyo) and one other commercial radio station in Tokyo (later TBS) were deferred for budgetary and other reasons. This decision was based on the principle of the co-existence of NHK and commercial stations in television broadcasting. The day Japan's first provisional television licence was granted marked the abolition of the Commission after two years and one month.

On 26 December 1952 the Minister of Posts and Telecommunications granted NHK a provisional licence for its Tokyo television station. NHK launched into construction work without delay, and on 1 February 1953 started Japan's first television broadcast service. NTV, the first commercial television station, began telecasting on 28 August the same year.

Both NHK and NTV tried to develop special kinds of

programmes, such as professional sports and other outside broadcasts. They also made efforts to popularize television by installing receiving sets on the streets and other places. However, the expense of television sets along with other factors limited the number of registered television viewers and as of the end of March 1954 there were only 17,000, compared with 11,709,000 radio listeners. In 1955 TBS appeared on the scene as the first commercial combined radio-television company.

Beginning in February 1956 the Ministry of Posts and Telecommunications began to expand the number of television channels and allocated these channels among major Japanese cities. On 21 May 1957 the Ministry adopted a basic policy for TV, stipulating that progress in commercial television was to advance in parallel with the realization of an NHK nationwide network, and also making the provision that more than one commercial television station would be able to operate in principal regions. This basic policy also made it clear that the establishment of education-oriented television stations was envisaged for the future, in order to meet the desires expressed by NHK and others. Based on this policy, on 19 June 1957 a plan was drawn up for the allocation of television channels in 50 regions throughout the country. This was known as the 'First Channel Plan'. Barring some changes made in the number of regions and channels, it is still in force.

On 22 October 1957 on the basis of the First Channel Plan (revised twice within that year), the Ministry of Posts and Telecommunications granted provisional licences to 36 commercial television stations of 34 companies, and to six more NHK television stations and an NHK educational television station in Tokyo. Beginning in June 1958, one after the other, the commercial stations went into operation. On 10 January, NHK also started the operation of its Educational Television channel in Tokyo and later opened exclusively educational television channels in other major Japanese cities as well. In 1958 receivers' contracts for television exceeded the 1 million mark.

On 24 December 1957 NHK inaugurated experimental FM broadcasting in Tokyo.

On 26 December 1957 the Ministry of Posts and Telecommunications issued provisional licences to NHK and NTV for the conduct of experimental colour television (formal licences were granted to both corporations the following day). Each of the two experimental stations adopted the NTSC colour system, and as a result, a Government decision in favour of the operation of Japanese colour television on the NTSC system was made on 18 June 1960. On 10 September 1960 NHK (Tokyo and Osaka) and four commercial colour television stations began regular colour telecasts under official licences. Along with the ever-increasing improvement of microwave circuits by the Japan Telegraph and Telephone Corporation, colour television broadcasting continued its expansion on a nationwide scale. The demand for colour television sets, however, did not show a corresponding increase during the following several years, hampered by the high price of the sets and the relatively short duration of colour telecasts.

Following the start of television broadcasting per se, the business of NHK expanded rapidly and commercial television stations also grew in number. In order to meet the new situation, on 15 March 1956 the Ministry of Posts and Telecommunications entrusted to the Special Broadcast Law Deliberation Council, composed of 15 citizens of learning and experience, the task of studying a draft bill for revision of the Broadcast Law. Upon receipt of the Council's report, the Ministry formulated a bill for amendment of the Broadcast Law on 13 July 1956. The bill, with some changes, was passed by the Diet on 13 March 1959, promulgated on 23 March and put into force on 22 April.

The amendments concerned the operations, organization and financial structure of NHK, and also required broadcasters in general to set up broadcast programme consultative organizations or to take other measures to maintain the appropriateness of their broadcast programmes.

In 1961, the Ministry of Posts and Telecommunications decided on the 'Second Channel Plan' for television, supplementing the first one, with the aim of encouraging the further spread of television by NHK and commercial television companies. By mid-1962, commercial television companies had succeeded in forming loose programming networks.

Contributing greatly to the spread of television sets among the people was the tremendous popular enthusiasm to be able to watch on television the wedding ceremony of Crown Prince Akihito and the gorgeous parades which followed (10 April 1959) and also the Tokyo Olympics which started on 10 September 1964. The number of receivers' contracts rose from 1,982,000 at the end of March 1959 to 4,149,000 at the end of March 1965.

Such tremendous growth in television audiences necessitated a revision in the contract fee structure and, in April 1962, a combined TV-radio charge and a radio-only charge were introduced.

(d) Investigation of overall aspects of broadcast legislation and recent developments in broadcasting

October 1962 saw the establishment of a Provisional Broadcast Legislation Investigation Council, a consultative organ for the Minister of Posts and Telecommunications, which was to be composed of citizens of learning and experience and was to devote itself to the study of overall aspects of legislation related to broadcasting. On 8 September 1964 the Council submitted a report of its two year research to the Minister. The report covered a very broad area, including the systematic utilization of radio waves, the solution of several problems regarding NHK and commercial broadcasters, continuing to maintain the dual system as before, maintaining appropriateness of broadcast programmes while ensuring freedom of programme compilation, and so forth. Based on this report, the Ministry picked up items requiring legislation to make them into law. On 15 March 1966 the Cabinet submitted to the Diet two related bills for amendment of the Radio Law and the Broadcast Law. Although a compromise between the Government and the Opposition was attempted concerning the bills, the Diet session closed before they could be passed.

Apart from this, in 1967, a partial amendment of the Broadcast Law allowed that, as of April 1968, owners of radio receivers alone would not have to pay the NHK receiver's fee.

Up to this time, both NHK and commercial television stations had gradually been increasing their colour broadcasts and in April 1968, NHK altered the contract fee structure to the present one.

Prior to this, the Ministry took steps to allow viewers throughout the country to have a choice of more than one commercial TV station. On 13 October 1967 a policy was adopted such that UHF channels, previously allocated only to 'Second Channel Plan' districts and other smaller regions, were to be allocated to 'First Channel Plan' districts as well, side by side with VHF channels.

In 1968 the Ministry of Posts and Telecommunications prepared technical and other standards in order to pave the way for full-scale FM broadcasting. On 19 November of that year, the Ministry determined its channel plan. Under this plan, in order that the special qualities of FM broadcast be made best use of, NHK was to direct its FM broadcasts to each prefecture, and commercial stations, for the time being, were allowed to operate in Tokyo, Osaka, Nagoya and Fukuoka. On 1 March 1969 a total of 170 NHK experimental stations aimed at the practical use of FM broadcasts (stereophonic stations were designated as experimental stations) started regular broadcasts, ending their experimental stage which had continued since 1963. On 24 December 1969 commercial FM broadcasting was inaugurated in Nagoya and, by May 1970, was operating in three other major areas too.

Chapter 3

PRESENT BROADCASTING STRUCTURE

1 LAWS AND REGULATIONS GOVERNING BROADCASTING

(a) Basic structures

Before World War II, telecommunications were, in principle, a government monopoly, and their use by the people or private entities was only permitted in exceptional cases. Broadcasting services, however, were sanctioned to private management from the start although strong government control was exercised over broadcasting both before and during the war.

The present broadcasting laws, enacted in 1950, were based on the philosophy that radio waves should be utilized by the people at large and that the way to achieve further advances was to initiate the principle of competition in the field of broadcasting. Thus, the dual system of NHK, which is a public-service corporation, and private (or commercial) broadcasters, which were to be private enterprises, came into being. This dual system may be said to constitute an outstanding feature of Japanese broadcasting laws.

Before explaining in concrete terms about Japanese laws related to broadcasting, it is necessary to give a brief outline of their basic structure. There are two fundamental laws, namely, the Radio Law which contains the general code on the use of radio waves, and the Broadcast Law which sets forth a specific code concerning broadcasting. In addition, there was the Radio Regulatory Commission Establishment Law, concerning the Radio Regulatory Commission which was to work as an independent regulatory commission controlling both radio and broadcasting administration. This law, enacted in 1950, was abolished in 1953.

The Radio Law aims to promote the public welfare by ensuring the equitable and efficient utilization of radio waves, including those for broadcasting, and, based on the ideal of their wide use by the public, attempts to regulate the technical aspects of use.

The Broadcast Law is intended to control broadcasting so as to meet the public welfare and to strive for the sound development thereof. It is for that purpose that the law provides for a dual system of NHK and commercial broadcasters, and imposes on NHK regulations concerning its organization, and business operation and also certain standards for its programmes. However, the law does not set any specific rules on the competency of a commercial broadcaster, but simply rules that any one, other than NHK, who has obtained a licence for a broadcasting station in accordance with the relevant provisions of the Radio Law becomes a commercial broadcaster.

The Radio Law was intended to exercise what might be likened to traffic control from a technical angle. In fact decisions on the use of a limited number of frequencies cannot be made only from a technical point of view. Since broadcasting services are expected to perform important social and cultural functions, frequency assignment or their licensing involves highly policy-oriented considerations. Although regulations on frequencies for broadcasting are treated much as those for other radio waves under the Radio Law, their actual application involves policy questions such as what kind of broadcasting services NHK and the commercial broadcasters should provide, which applicants for commercial broadcasting stations should be licensed, etc. Such considerations, however, were regarded as lying beyond the scope of this law when it went into effect.

Hereafter, we will give an outline of regulations on broadcasting stations under the Radio Law, and specific regulations regarding broadcasting under the Broadcast Law and other laws, due heed being paid to the above mentioned points.

As seen from the above, an important part of Japan's broadcasting administration comes under the Radio Law which regulates broadcasting stations as radio stations. In this respect, voices have been raised for the readjustment of the Broadcast and Radio Laws so that broadcasting administration is separated from telecommunications administration. The Government, in response to this demand, once presented bills for amendment of the two laws along those lines to the Diet. They have failed to pass the Diet, however.

(b) Regulations governing broadcasting stations

(1) Licensing of broadcasting stations

(i) Broadcasting frequency assignment plans: In accordance with the Radio Regulations annexed to the International Telecommunication Convention, the Minister of Posts and Telecommunications has charge of the assignment to various services of frequencies usable in Japan. The Minister's decisions made in this regard are promulgated in the 'Principle of Frequency Assignment'.

With regard to frequencies for broadcasting use, the 'Broadcasting Frequency Assignment Plans' provide for medium wave broadcasting, very high frequency FM broadcasting (sound), and television broadcasting. (No assignment plan has been drawn up for short-wave broadcasting. NHK conducts overseas shortwave broadcasts, while one commercial broadcasting company operates domestic shortwave broadcasts.)

These frequency assignment plans form the basis on which licences are granted to broadcasting stations. In this way, the formulation of the plans has played a very important role in the growth of Japan's broadcasting system. All major national policies on broadcasting in the past - for example, those which dealt with the start of television broadcasting, the number of channels allocated to NHK and commercial television stations, and the widespread diffusion of television broadcasting- have taken shape through the formulation of or amendments to frequency assignment plans.

However, it should be noted that no procedure for the formulation of 'Principles of Frequency Assignment' to various services is stipulated in the Radio Law, nor are there any legal grounds clarified for the formulation of broadcasting frequency assignment plans. Rather, these matters have been left to the discretion of the Ministry of Posts and Telecommunications, which frames its own policy regarding the issue of licences for broadcasting stations. This has become one of the controversial issues of the laws affecting broadcasting.

In formulating (or amending) the broadcasting frequency assignment plan, the Ministry of Posts and Telecommunications seeks opinions from those concerned by holding hearings or the like, and then presents a draft plan to the Radio Regulatory Council, a consultative body for the Minister of Posts and Telecommunications. The Ministry reaches a final decision, taking the opinion of the Council into consideration. This procedure is not specified by law, but has now become a sort of unwritten administrative rule. Under present frequency assignment plans, NHK is assigned frequencies for two national networks of medium-wave broadcasting, whereas commercial broadcasters are allotted frequencies to establish two or three stations in major metropolitan areas and one station in each prefecture in other regions. In the field of very high frequency (VHF-FM) broadcasting, NHK is allotted frequencies to establish a national network and commercial broadcasters frequencies to set up one station each in Tokyo, Osaka, Nagoya and Fukuoka, and none in other cities. In the field of television, NHK is allotted frequencies to set up two national networks: General Television (GTV) and Educational Television (ETV). Frequencies to set up four or five commercial broadcasting stations in major metropolitan areas and 2 stations each in other prefectures are also provided for.

(ii) Fundamental standards for the establishment of broadcasting stations: Any broadcasting station, in so far as it is a type of wireless station, is unable to open for service without obtaining a licence from the Minister of Posts and Telecommunications (Article 4, Radio Law). In reviewing applications for licence, the Minister uses the criteria stipulated in the Radio Law: conformity of facilities to the technical standards, availability of frequencies to be assigned, and a sound financial basis for maintaining business activities. In addition, the applicants must conform with the fundamental standards for the establishment of broadcasting stations prescribed in the Ministry of Posts and Telecommunications Ordinance (Article 7, Radio Law).

Incompetency for operating wireless stations is also covered in the Radio Law. The regulations concerning broadcasting stations are more strict than those for other wireless stations with such rulings as the exclusion of broadcasters of foreign origin (Article 5, Paragraph 4, Radio Law).

The 'Fundamental Standards for Establishment of

Broadcasting Stations' contains several points worthy of note.

First, the standards provide that the programmes broadcast should conform with the items corresponding to the programme codes as stipulated in the Broadcast Law (Article 44, paragraph 3, etc.) and that broadcasters prove conformity through past achievements at the time of licence renewal.

Second, the standards contain a general clause under which the establishment of the station should 'contribute to impartial and efficient dissemination of broadcasting'. The detailed standards for judgment of applications for licences in relation to this clause are set by the Ministry of Posts and Telecommunications and contain some very important points concerning, in particular, the granting of licences to commercial broadcasters. These detailed standards stress (a) a close relationship to be maintained by commercial broadcasters with the local community; (b) the prevention of monopolization of plural stations by any one person; (c) the exclusion of concentration of mass communication media in local communities (the limitation of ownership or participation in management of radio, television or newspaper companies by any one person, etc.).

Third, the standards invest the Minister of Posts and Telecommunications with broad powers to decide which applicants should be granted licences.

(iii) Licensing procedures: In cases where the application for licence meets the requirements stipulated in the law and the standards, the Minister issues a provisional licence or construction permit (Article 8, Radio Law). The recipient of the provisional licence is granted a full licence when the construction work is completed and is found to come up to the technical standard (Article 12, Radio Law).

In that this provisional licence is, in effect, tantamount to a full licence, its disposition becomes a matter for reference to the Radio Regulatory Council (Article 99-11, Radio Law).

It is stipulated that the effective term of licence for a broadcasting station should not exceed three years (Article 13, Radio Law). At present, the effective term of licence for new broadcasting stations is individually determined so as to make possible simultaneous re-licensing with existing stations whose effective term of licence has continued for three years.

The Radio Law provides that it is permissible to fix conditions or terms on either provisional or full licences (Article 104-2, Radio Law). At present, all television stations are required to present a certain percentage of educational and cultural programmes, and VHF-FM stations are required to present a certain percentage of stereophonic broadcast programmes.

(2) Operation of broadcasting stations

In addition to the special regulations concerning broadcasting as provided for in the Broadcast Law, broadcasting stations are required to observe the rules which are applied to radio stations in general under the Radio Law. These rules include the prohibition of use of station for other purposes (the prohibition of operation outside the realm of matters to be broadcast as defined in the licence, etc.), the prevention of interference, and so on. Any person found guilty of such acts as the despatch of false messages for the benefit of himself or any other person or to inflict damage on any other person, or of advocating the violent overthrow of the Japanese Constitution or the Government established thereunder, or of broadcasting indecent matter shall be subject to punishment.

(3) Supervision of broadcasting stations

Broadcasting stations, in so far as they are a type of radio station, are subject to supervision by the Minister of Posts and Telecommunications. Provisions of the Radio Law include enforced changes in assigned frequencies, etc. for the public interest, enforced inspection of radio equipment, and government orders for the conducting of communications for the saving of lives, etc. in times of emergency (damages caused or pecuniary losses incurred as the result of frequency changes or emergency communications may be compensated for by the Government). There are also rules for the suspension of operations or the revocation of licences in case of violation of the Radio Law or the Broadcasting Law or orders issued under those laws.

Although there have been no instances of the revocation

of licence for a broadcasting station to date, these rules have serious implications for a broadcaster, since the revocation of the licence for any part or all of his broadcasting activities would be disastrous. Furthermore, if these rules should be invoked in cases of violations concerning the programme code stipulated in the Broadcast Law (of which the content will be referred to later), the legality for such an invocation would be subject to considerable controversy, since such matters concern editorial freedom.

Regulations governing broadcasting services

(1) Basic purpose of the Broadcast Law

The Broadcast Law is intended to regulate broadcasting so as to meet the public welfare and to strive for the sound development thereof. For this purpose, the basic principles are set forth in the following manner:

(i) To secure the maximum availability and benefits of broadcasting to the people (Article 1, Item 1);
(ii) To assure the freedom of expression through broadcasting by guaranteeing the impartiality, integrity and autonomy of broadcasting (Article 1, Item 2);
(iii) To make broadcasting contribute to the development of healthy democracy by clarifying the responsibility of those persons engaged in broadcasting (Article 1, Item 3).

Based on these principles, the Broadcast Law sets forth various regulations in concrete terms.

First, under the Broadcast Law, which covers a dual system of public (NHK) and private (commercial) broadcasting the people are meant to have the maximum availability to and benefit from broadcasting.

Second, the Broadcast Law enshrines freedom of expression through broadcasting by stipulating that 'Broadcast programmes shall never be interfered with or regulated by any person, excepting in cases where it is done with powers provided for by law' (Article 3). This law ensures the autonomy of broadcasting and, at the same time, requires broadcasters to voluntarily adopt broadcast programme standards, etc. for the sake of guaranteeing the impartiality and integrity of broadcasting.

Third, the Broadcast Law clarifies the powers and responsibility of the Board of Governors and other organs of NHK, and sets down the responsibilities of NHK and commercial broadcasters with regard to the compilation and broadcasting of broadcast programmes.

(2) Dual system of broadcasting

(i) NHK: The NHK is neither a state-owned broadcasting organization funded from national revenue nor a broadcasting organization financed from advertising. It is a non-commercial public corporation funded almost entirely by receivers' fees paid by the viewers, and required to provide a nation-wide service.

The Broadcast Law makes careful provision for the autonomy of NHK management so that its business operation is worthy of the nature of a people's broadcasting organization, and that the wishes of the people are reflected adequately.

First, the Broadcast Law provides for the establishment of a Board of Governors as the highest decision-making organ of NHK. The Board is composed of 12 Governors appointed by the Prime Minister with the consent of both Houses of the Diet. Its duties include decision on management policy and other important matters, the appointment or dismissal of the President, who heads the executive organ of NHK, and of auditors, the auditing organ, and approval of the appointment by the President of the Vice President and Managing Directors.

Second, the Broadcast Law provides for a system of receivers' fees, by which system NHK is authorized to collect receivers' fees on the basis of receivers' contracts concluded with the receivers, in order to finance the maintenance and operation of its businesses. Aside from this, it is stipulated that NHK shall not broadcast any advertisement related to any other person's commercial business (Article 46). Consequently, NHK has no revenue whatever accruing from advertising.

Third, the Broadcast Law provides in concrete terms for NHK's activities including domestic broadcasting, research and investigation into broadcasting, and international broadcasting services - all of which NHK is required to execute under this law - and also, other operations to be conducted when deemed necessary. In this respect, the law places restrictions upon NHK by forbidding any intention of profit-making (Articles 9, 9-2, 39, etc.). With regard to domestic broadcasting, the law further provides that NHK shall 'broadcast rich and good broadcast programmes', 'have local programmes as well as national programmes' and, that 'it shall strive to be conducive to the development and popularization of new civilization as well as to the preservation of the past excellent civilization of our country'. Also, it requires NHK to conduct scientific listening polls at regular intervals in order to know the wishes of the people concerning programming in domestic broadcasting (Article 44, Item 1, and 44, Item 2).

Fourth, the Broadcast Law provides for public control to be exercised over NHK through the Government and the National Diet.

For instance, it is provided that NHK shall prepare the budget of revenues and expenditures, business projects and financial programmes for every financial year and present them to the Minister of Posts and Telecommunications, and the Minister shall transmit them, with his comments, through the Cabinet to the Diet for approval. The monthly fee to be collected from a person who has concluded a receiver's contract, according to the stipulation, shall be determined by the Diet's approval of the budget of revenues and expenditures as prepared by NHK (Article 37). Similarly, the settlement of accounts of NHK is presented to the Diet (Article 40).

The financial report is also submitted to the Diet, with the comments of the Minister of Posts and Telecommunications (Article 38). The prominent feature of public controls over NHK is that all controls are exercised primarily by the Diet which represents the people, and that the executive branch of the Government plays a minimal role.

(ii) Commercial broadcasters: The Broadcast Law provides that commercial broadcasters are those who have obtained licences under the Radio Law (excepting NHK). It simply stipulates a few regulations such as that the broadcast programme standards required of NHK be applied to commercial broadcasters. Therefore, it may safely be said that the law envisages commercial broadcasters as operating freely and without restrictions.

The noteworthy points in the regulations governing

commercial broadcasters are that the administrative policy regarding the granting of licences under the Radio Law has stressed the close relationships with the local community and the restriction placed on the domination of plural stations by any single person or company. This policy seems to have been motivated by the consideration that commercial broadcasting, aided by its close relationship with local communities and its presentation of local characteristics, should contribute to the sound development of Japanese broadcasting side by side with NHK which, under the Broadcast Law, aims at national broadcasting over its national networks.

In this regard, the Broadcast Law rules that commercial broadcasters must not conclude any agreement relating to the supply of broadcast programmes which includes any terms on which broadcast programmes are supplied by a particular person only (Article 52-3). As a matter of fact, however, commercial broadcasters - commercial telecasters in particular - have found it very difficult to fill their programming schedule entirely with programmes of their own, leading them to form virtual networks within the limitations of the Broadcast Law.

The Broadcast Law presupposes that commercial broadcasters obtain income from advertisements, and requires them to announce in the broadcast that it is a sponsored broadcast (Article 51-3). The law also provides that the educational programmes intended for schools must not include any advertisement obstructive to school education (Article 52-2).

(3) Regulations of broadcast programmes

The Broadcast Law ensures the freedom of programme compilation (Articles 1 and 3) while establishing standards regarding the production of domestic broadcast programmes in addition to the rule governing NHK only (Article 44, paragraphs 1 and 2), so that the broadcasters may secure the appropriateness of their programming on their own initiative. In this connection, the law sets forth several rules.

(i) The law includes the following four standards applicable to the programmes of both NHK and commercial broadcasters (Article 44, paragraph 3 and Article 51). Programmes:

Shall not disturb the public security and good morals and manners;

Shall be politically impartial;

Shall broadcast news without distorting facts;

As regards controversial issues, shall clarify the point of issue from all angles possible.

In compiling domestic broadcast programmes, the broadcasters are required to provide cultural programmes or educational programmes as well as news programmes and entertainment programmes, maintaining harmony among the broadcast programmes (Article 44 paragraph 4, Article 51). Furthermore in compiling educational programmes, the broadcasters are required to clearly indicate the persons at whom the broadcasting is aimed, to make the contents of the broadcasting systematic and continuous as well as instructive and appropriate to such persons; to allow the general public to learn the plans and the contents of the broadcasting in advance; and, if the programme is intended for schools, to make the contents conform to the standards of the curricula provided for by the laws and regulations relative to the school educational system (Article 44 paragraph 5, Article 51).

(ii) The Broadcast Law provides that the broadcasters shall establish standards for the compilation of broadcast programmes (Article 44-2, Article 51), and have Broadcast Programme Consultative Committees in order to maintain the appropriateness of the broadcast programmes (Article 44-3, Article 44-4, Article 51-2).

In relation to the guarantee of the integrity of broadcasting (Article 1) and the programme standards for news broadcasting without distortion of the facts (Article 44, paragraph 3, item 3), the Broadcast Law stipulates that, upon demand from a person whose rights have been infringed upon by broadcasting or any person concerned in the case by reason of the contention that a broadcaster broadcast some untrue matters, must investigate whether such broadcast matter was true or not and if the objection is upheld must broadcast the correction or withdrawal on certain conditions (Article 4). The broadcasters are required to ensure that the Broadcast Programme Consultative Committees or the plaintiffs can verify the contents of the programme(s) in question within three weeks of the broadcast (Article 44-7, Article 53).

(iii) Article 3 of the Broadcast Law sets down: 'Broadcast programmes shall never be interfered with or regulated by any person, excepting the case where it is done by the powers provided for by law.' In addition to these provisions, programming is regulated under other laws.

First, the Broadcast Law requires impartial treatment of candidates for elective office in the broadcast of their political views (Articles 45 and 52). The Public Office Election Law provides that the broadcasters shall broadcast the political views or information about the careers of the candidates in the elections of both Houses of the Diet or gubernatorial elections. In this case, the broadcasters, including NHK, have to make the broadcasts free of charge, either by video or sound prerecording (Article 150, the Public Office Election Law). In addition, the broadcasters are subject to regulations different from those governing newspapers and magazines. These ensure impartial election broadcasts (Article 151-3 the Public Office Election Law), and restrictions on election campaign broadcasts (Article 151-5 the Public Office Election Law).

Second, the Radio Law empowers the Minister of Posts and Telecommunications to order such communications as are found necessary for the saving of lives or for the protection of traffic and communication facilities at time of emergencies, such as earthquakes, typhoons, floods, tidal waves, etc. (the state is obliged to compensate for the actual expenses needed for such communications) (Article 74).

The Fundamental Law of the Measures Against Calamity names NHK as one of the designated public organs which are required to draw up and execute a programme for the prevention of disasters, and thereby contribute to that prevention (Articles 2 and 6). At the time of extraordinary natural phenomena such as storms, heavy rain, heavy snow, floods, flood tides, earthquakes, tidal waves, conflagration or explosion, relevant authorities can ask NHK and commercial broadcasters to broadcast requests and warnings (Article 57, the Fundamental Law of the Measures Against Calamity).

In addition, broadcasting stations are required to perform many duties for co-operation with the work of dealing with disasters. The Weather Service Law and the Flood Defence Law can be cited as examples. The Weather Service Law, Article 15, provides that the Meteorological Agency, when it has issued an alarm on certain weather - tidal waves, flood tides, high seas and floods - should immediately inform NHK of the content, and NHK is required to broadcast the item(s) without delay. The Flood Defence Law, Article 10, provides that the Director General of the Meteorological Agency, upon discovery of any sign of floods or flood tides under certain weather conditions, is bound to seek, if necessary, the co-operation of the broadcasting stations, newspaper companies, news agencies and other information organs and to have them inform the general public of the situation. The Construction Minister also is empowered to do the same in cases when it is feared that rivers of certain classes may flood.

The Japan Red Cross Society Law provides that the Society be provided with facilities and conveniences relating to its communications on relief work, so that appropriate relief work can be carried out promptly (Article 34, paragraph 2).

Third, obscene actions, commodities, publications, etc., as provided for in the provisions concerning morals in the Criminal Code (Article 174) and the provisions of the Radio Law concerning obscene communications (Article 106), are prohibited. Any person who has conducted any communication advocating the violent overthrow of the Japanese Constitution or the government established thereunder, is subject to punishment (Article 107, Radio Law).

The broadcast of advertisements is subject to regulations governing advertisements in general. For instance, misleading representation is prohibited by the Act against Unjustifiable Premium and Misleading Representation, and exaggerated advertisements are also banned. Under the Medicine Law, the advertising of medicines for specific diseases is subject to certain restrictions and, pending government approval, is banned. The Food Hygiene Law provides for the prohibition of untrue representation, etc. and the Petty Offence Law punishes the act of making an advertisement which asserts a deceiving or misleading fact.

Finally, reference is made to the relationships between the broadcast services and the Copyright Law.

The Copyright Law in Japan was totally revised in 1970, and under the new law, various measures for the protection of copyright holders have been taken. These include the neighbouring rights, reproduction rights, re-broadcasting rights, cable broadcasts rights, television distribution rights, etc. (cf. Article 13, Cable Television Broadcast Law (re-transmission), Article 5, Law to Regulate the Operation of the Cable Sound Broadcasting service (consent of re-transmission)).

(d) Regulation of cable broadcasts

(1) Regulation of cable sound broadcasting service

The cable sound broadcasting service in Japan originated from the installation of community listening facilities in rural areas without good reception. Beginning in 1950 receiving equipment with microphone attached community listening facilities came into increasing use for the distribution of information in limited local areas. In the latter part of the 1950s, handsets were affixed to these facilities, leading to the cable sound broadcast-telephone service being utilized also as a telephone service among subscribers.

In addition, in urban areas, street-advertising was conducted with the use of cable sound broadcasting equipment, which was used for transmitting music to coffee shops and restaurants.

In 1951, the Law to Regulate Operation of Cable Broadcasting Services was enacted, and any person who wished to start such a service was required to notify the Minister of Posts and Telecommunications in advance. Also, it was ruled that the freedom of programme compilation (Article 3) and the standards for broadcast programmes (Article 44, paragraph 3) and other provisions of the Broadcast Law were to be applied to the programmes of the cable radio broadcast service. It was also made clear that those responsible for this service should obtain the consent of the broadcasters concerned, with regard to the re-transmission of broadcast programmes.

The Law was amended in 1973, at the time when the Cable Television Broadcast Law was enacted, and came to be known as the Law to Regulate Operation of Cable Sound Broadcasting Services, and defined this service clearly.

With regard to cable broadcast-telephone services operated by means of cable sound broadcasting facilities with handsets attached, in 1957 a law was passed which stipulated that the service be subject to permission from the Minister of Posts and Telecommunications.

In addition to these laws, there is the Cable Telecommunications Law governing cable telecommunications in general. Under this law the authorities must be informed of the installation of necessary equipment, the equipment must conform to certain technical standards, and linkage with other services is restricted. These provisions are naturally applied to the cable sound broadcasting service and also to the cable television broadcasting service, to the latter of which reference follows.

(2) Regulations of Cable Television Broadcasting Service

Cable television started in 1955, two years after the start of television broadcasts, with NHK experiments on community reception facilities. Later, when there were found to be areas still suffering from poor reception in the mountainous regions, despite the increased establishment of relay-broadcast stations, community reception facilities were increased even further.

NHK, charged with the nationwide provision of broadcasting services, began providing financial assistance for constructing such facilities in 1960, and in 1971 started joint construction and operation work on these facilities with the local viewers. The greater part of the existing cable television broadcasting service in this country is occupied by television community-reception facilities. 43 per cent are jointly owned and operated by NHK and local viewers.

These facilities much improved reception in mountainous areas, and so when similar problems resulting from the increased number of high-rise buildings began to be experienced in urban areas, major metropolitan areas including Tokyo began to construct like facilities. From 1965 with the development of cable television services in the USA as profit-making enterprises, strong arguments came to be heard that by combining the service of re-transmission of broadcast programmes and that of transmitting original programmes, these services could become an influential part of mass communication. In 1968 a commercial service which included cable television broadcasts among its operations, was established in Tokyo.

Up to this time, the Ministry of Posts and Telecommunications had been applying the Law to Regulate the Cable

Broadcasting Service to the cable television broadcasting service as well, requiring notification of the start of service, etc. But, prompted by the fear that a possible increase in the number of cable television broadcasting services established on a profit-making basis and without restriction would be harmful to the public interest, the Ministry presented a bill to the Diet for amendment of the Law in 1969 which included a provision that would have required a re-transmission service for television programmes broadcast in urban areas to obtain government permission. This bill was shelved and withdrawn. In 1970, however, the Ministry exercised administrative guidance by securing the participation and collaboration of broadcasters and others and made possible the establishment of a public utility corporation in Tokyo designed for the operation of a re-transmission service of television broadcast programmes. Similar corporations came into being in the Kyoto-Osaka-Kobe area, the Nagoya area and the Fukuoka area.

Popular demand was not yet satisfied, however, A series series of plans for the establishment of cable television in all areas were drafted. To cope with the situation, the Ministry of Posts and Telecommunications submitted a Cable Television Broadcast Bill to the Diet in 1971. This bill, which included the provision that any service reaching beyond a certain limit should obtain government permission, was passed with partial amendment in June 1972.

The Cable Television Broadcast Law went into force in January 1973 and is still in effect. The salient provisions of this Law are:

(i) Permit for installation of facilities: The establishment of cable television broadcasting facilities of a scale exceeding a certain limit with the aim of conducting a cable television broadcasting service requires a permit from the Minister of Posts and Telecommunications. The recipient of this permit should, in principle, allow the use of facilities to those who have requested transmission of their programmes.

(ii) Notification on opening of operation: A person who intends to become a cable television broadcast operator is required to notify the Minister of Posts and Telecommunications of the service area, re-transmission service and other specified matters. The operator cannot refuse service within the service area without appropriate reason. Provisions of the Broadcast Law cover programming.

Within any area experiencing reception difficulties and designated as such by the Minister, the simultaneous re-transmission of television broadcast programmes is obligatory. No such area has yet been designated, however.

(iii) Cable Broadcast Council: An amendment made in the Diet provided for the establishment of a Cable Broadcast Council attached to the Ministry of Posts and Telecommunications. This Council advises the Minister on the question of permission regarding the installation of facilities and other matters and may give recommendations on matters considered necessary. The Council is composed of seven part-time members appointed by the Minister of Posts and Telecommunications.

(iv) Other matters: In addition, there are provisions covering Ministerial procedures for improvement of facilities and service, the revocation of permits, the order for suspension of services, penalty against the installment of illegal facilities, and so forth.

The law provides that facilities needing government permission are those equipped with 501 or more incoming terminals, and that those equipped with 50 or less incoming terminals need not notify the commencement of operation. Facilities with 50 or less terminals must, however, give notification to the Minister if they are to originate programmes. The purpose of this law is to protect the interests of the audience of cable television broadcasting and to strive for the sound development of cable television broadcasting by regulating the establishment of the facilities thereof and the operation of the service thereof so as to contribute to the promotion of public welfare. The Minister of Posts and Telecommunications, when granting a permit, must be satisfied that the plan of the cable television broadcasting facilities is reasonable and its practicability is ensured; that the cable television broadcasting facilities satisfy the technical standards prescribed by the Ministry of Posts and Telecommunications Ordinance; that there is present the financial basis and technical ability sufficient for ensuring the establishment and operation of the cable television broadcasting facilities; that the establishment of such facilities is needful and appropriate; and that the Minister of Posts and Telecommunications consult the relevant prefectures before granting a permit.

2 BROADCASTING STRUCTURE

(a) Establishment and abolition of Radio Regulatory Commission

In 1950, when the present broadcasting system came into being, the Radio Law, the Broadcast Law and the Radio Regulatory Commission Establishment Law were formulated. Based on the Radio Regulatory Commission Establishment Law, the Radio Regulatory Commission was established in June that year.

The Allied Occupation in Japan saw a vigorous introduction of the American type of administrative system along with a large number of newly established administrative commissions based on the American model. The Radio Regulatory Commission was one of these, established as an independent regulatory commission vested with overall powers in the area of regulations relating to radio waves and broadcasting. The Radio Regulatory Commission composed of one chairman and six members, was empowered to formulate Radio Regulatory Commission Ordinances, thereby exercising a quasi-legislative function.

It was responsible for several important decisions directly related to the development of Japan's broadcasting under the new post war broadcasting set up such as the approval of licensing for commercial radio broadcasting in 1951 (giving rise to commercial stations), the decision on the standard system for television broadcasting in 1952, and the formulation of a policy for licensing television broadcasters.

The Commission, as with other administrative commissions which were dissolved after the San Francisco peace treaty went into force, was abolished in July 1952. Thereafter the Ministry of Posts and Telecommunications had the final decision on matters regarding radio and broadcasting administration.

(b) The Ministry of Posts and Telecommunications

The Ministry of Posts and Telecommunications is in charge of regulations concerning radio, broadcast and cable broadcast. The Ministry conducts other operations such as the national postal service and a postal savings service, is charged with regulating cable telecommunications services and is vested with powers which cover a broad range of communication administration.

With regard to radio and broadcasting administration, the Radio Regulatory Consultative Council attached to the Ministry plays an important role. In formulating Ministerial Ordinances under the Radio Law, granting provisional licences, or giving permission under the Broadcast Law, the Minister is required to refer the issue to the Radio Regulatory Consultative Council, and take action in accord with the decision of the council.

Any objections raised to a provisional licence or any other dispositions made under the Radio Law are required to be submitted to the Council, which, after quasi-judicial procedures including the holding of hearings, etc., will vote on a finding. The Minister must act on the decision of the council's finding.

The Council is composed of five part-time members who are appointed by the Minister of Posts and Telecommunications with the consent of both Houses of the Diet.

It is sometimes argued that the Council should be invested with greater powers and that it should execute a more impartial, neutral administration with regard to the regulations concerning radio and broadcasting. Similarly, the opinion that an independent regulatory commission should be established in the field of broadcasting administration is also heard. The administrative structure of broadcasting in Japan poses a problem for future solution.

Within the Ministry of Posts and Telecommunications, the Radio Regulatory Bureau has charge of activities concerned with radio and broadcast administration. There are also regional radio regulatory bureaux which work as local branches.

3 EXISTING STATE OF BROADCASTING SERVICES

(a) Number of broadcasting stations

The number of broadcasting stations operated by NHK plus 107 commercial broadcasters is 8,274. Of these, 491 are medium wave stations, 3 shortwave stations, 471 VHF FM stations, and 7,309 TV stations. The 107 commercial broadcasters account for 54 TV, 17 radio, and 36 radio-TV companies.

TABLE 1 NHK and Commercial Stations (as of 31 March 1977)

Classification	NHK		Commercial stations		Total
	Class	No. stations	No. companies	No. stations	
MW broadcasting	1st Network	173 (119)			
	2nd Network	141 (90)	48	177 (129)	491 (338)
	TOTAL	314 (209)			
SW broadcasting	Domestic	-	1	2 (1)	
	International	1	-	-	3 (1)
	TOTAL	1	1	2 (1)	
VHF FM broadcasting	-	464 (406)	4	7 (3)	471 (409)
TV	General	2,495 (2,443)			
	Educational	2,452 (2,407)	90	2,362 (2,272)	7,309 (7,122)
	TOTAL	4,947 (4,850)			
GRAND TOTAL		5,726 (5,465)	-	2,548 (2,405)	8,274 (7,870)

( ) indicates the number of relay stations.

(b) Broadcast programmes

NHK's broadcasting systems include General Television, which presents a harmonious combination of news, educational, cultural and entertainment programmes for the general public, and ETV which devotes itself to educational programming, and a limited amount of cultural and news programmes. The First Radio Network is directed towards the general public, presenting news, education, culture and entertainment, the Second Radio Network presents education-oriented national programming, and VHF broadcasting is devoted to news and informational programming plus programmes highly characteristic of FM broadcasting. On the other hand, local programming is prepared on a prefecture-to-prefecture basis, while General TV constitutes the mainstay of television broadcasting, and radio broadcasting is mainly fed through the First Radio Network and VHF broadcasts.

Commercial TV broadcasting provides well-balanced general programming including educational, cultural and entertainment programmes. There are from two to six such commercial stations in each region, keenly competing for high audience ratings and, consequently, inclined to concentrate on entertainment. Television programmes are expensive, and the fact that a large amount of talent is concentrated in the Tokyo area has given rise to several national networks whose key stations in Tokyo provide the bulk of commercial programming. Commercial radio broadcasting on the other hand, is much more localized, with emphasis on the presentation of local information interwoven with music. However, most news and sports programmes of national importance are broadcast over nationanetworks. Commercial FM broadcasting is required by law to present 70 per cent or more of its programmes in stereophonic broadcasting and therefore its programmes are mostly musical.

In 1969, NHK and the National Association of Commercial Broadcasters in Japan joined hands in establishing the Broadcaster's Council for Better Programming. In addition, in 1968, the Broadcast Programming Center of Japan was set up with the object of improving and enriching educational and cultural commercial programming, the starting fund as well as the yearly operation costs being provided by NHK and commercial broadcasters.

The average daily broadcasting hours of NHK and commercial broadcasters are shown in Table 2. Broadcasting hours classified by programme category are given in Tables 3a and 3b.

TABLE 2 Average daily broadcasting hours

NHK (Tokyo)		
GTV	17h 44	(Colour: 17h 42)
ETV	18h 00	(Colour: 12h 50)
R1	19h 05	
R2	18h 30	
VHF FM	18h 02	(Stereophonic: 12h 46)
Commercial stations*		
TV	16h 52	
Radio	21h 23	

(*These figures represent an average taken from 53 MW, SW and VHF FM commercial radio stations and 90 TV companies)

(c) Television news

As indicated in Table 4 and 5 television plays a prominent role in enabling the public to acquire news and enrich their knowledge and culture.

(d) Listening and viewing habits

The TV viewing hours of the public have been showing a steady increase in recent years, and radio listening hours a slight increase, too.

The percentage of persons actually listening and viewing in the period 1972-6 is given in Table 6 and the average listening and viewing hours are given in Table 7.

(e) Registered NHK Viewers

The number of NHK's receivers' contracts as of 31 March 1977 is as follows, per household.

Ordinary contracts	Colour contracts	Total
3,749,433	23,309,448	27,058,881

Ordinary contract means receivers' contracts excepting colour TV. Colour contract means receivers' contracts including colour TV.

TABLE 3a Radio-TV broadcasting hours by categories, NHK 1976-7

Categories	CTV			ETV		
	Hrs	Ratio %	Weekly average	Hrs	Ratio %	Weekly average
News	2,376h 18	36.7	45h 34	54h 55	10.8	1h 03
Education	1,025h 21	15.9	19h 40	5,167h 10	78.7	99h 06
Cultural	1,585h 03	24.5	30h 24	1,347h 55	20.5	25h 51
Enter-tainment	1,483h 39	22.9	28h 27			
TOTAL	6,470h 21	100.0	124h 05	6,570h 00	100.0	126h 00

Categories	Radio 1			Radio 2		
	Hrs	Ratio %	Weekly average	Hrs	Ratio %	Weekly average
News	2,906h 38	41.7	55h 44	730h 25	10.8	14h 01
Education	224h 54	3.2	4h 19	5,171h 10	76.6	99h 10
Cultural	2,115h 57	30.4	40h 35	850h 55	12.6	16h 19
Entertain-ment	1,718h 41	24.7	32h 58			
TOTAL	6,966h 10	100.0	133h 36	6,752h 30	100.0	129h 30

TABLE 3a (continued)

Category	FM		
	Hrs	Ratio %	Weekly average
News	926h 49	14.1	17h 46
Education	438h 00	6.6	8h 24
Cultural	3,416h 29	51.9	65h 31
Entertainment	1,801h 17	27.4	34h 33
TOTAL	6,582h 35	100.0	126h 14

Cable broadcasting services (as of 31 March 1977)

(1) Cable radio

Cable radio facilities number 7,732 of the total, those conducting cable radio broadcasting alone number 4,628, those combining broadcasting with telephone services 1,111, those conducting advertisement broadcasting on the streets 1,464 and those services presenting musical broadcasts 529.

(2) Cable television

Cable television services are divided into 9,986 small scale ones which need only report to the authorities at the time of inauguration, and 181 larger scale services which are required to obtain government authorization at the time of establishment, bringing the total to 10,167. Most of these services are intended to surmount viewing difficulties in remote regions or viewing disturbances in urban areas. They mainly re-transmit broadcast programmes. NHK operates community reception services in 7,054 areas in remote regions.

TABLE 3b Radio-TV broadcasting hours by categories, commercial broadcasting, 1976-7

Category	TV				Radio			
	Sponsored		Self-sustaining		Sponsored		Self-sustaining	
	Hrs	Ratio %	Hrs	Ratio %	Hrs	Ratio %	Hrs	Ratio %
News	202h 11	11.2	109h 39	14.5	823h 36	14.0	203h 12	10.6
Education	655h 12	12.1	103h 54	13.8	334h 36	5.7	66h 15	3.5
Cultural	1,365h 42	25.3	159h 24	21.2	1,190h 36	20.2	322h 30	17.0
Entertainment	2,535h 27	47.0	330h 30	43.9	768h 54	13.0	316h 36	16.6
Music					2,255h 36	38.3	955h 0	50.0
Sports	196h 21	3.6	17h 51	2.4	442h 6	7.5	33h 57	1.8
Commercials	31h 51	0.6	4h 33	0.6	60h 51	1.0	2h 30	0.1
Others	8h 18	0.2	27h 6	3.6	17h 24	0.3	7h 51	0.4
TOTAL	5,399h 24	100.0	725h 57	100.0	5,896h 39	100.0	907h 51	100.0
Percent	87.8%		12.2%		75.6%		24.4%	100%

TABLE 4 News sources (people of 13 years or over; multiple replies)

	Daily news	Major events	Emergency news
	%	%	%
NHK TV	81.6	85.8	83.7
Commercial TV	47.4	42.8	41.8
National papers	46.5	41.5	29.2
Local papers	29.3	24.0	17.5
NHK radio	13.0	9.1	22.1
Commercial radio	12.4	7.8	14.6
Others	1.3	1.2	1.4

(Source: NHK's National Survey on People's Inclinations, May 1975)

TABLE 5 Media utilization for enrichment of knowledge and culture (people of 13 years or over; multiple replies)

	Common knowledge and culture	Practical knowledge and techniques	Life philosophy
	%	%	%
NHK TV	54.8	38.5	30.5
Commercial TV	29.4	21.4	17.1
NHK radio	8.1	5.1	4.8
Commercial radio	7.3	4.4	4.3
Newspapers	60.1	36.1	31.7
Magazines	34.2	32.7	21.6
Books	34.0	34.8	38.5
Dialogues	35.6	24.6	41.0
Films and concerts	9.8	5.3	7.0
Lectures, classrooms	11.3	15.8	11.4
Others	0.7	0.4	1.0

(Source: NHK's National Survey on People's Inclincations, May 1975)

TABLE 6 Changes in actual listening and viewing persons

Survey dates		1972		1973		1974		1975		1976	
		June	Oct.	June	Nov.	May	Nov.	June	Nov.	June	Nov.
		%	%	%	%	%	%	%	%	%	%
TV	Week-days	94	95	95	95	92	93	92	93	92	91
	Sun-days	93	94	93	94	93	96	92	95	93	94
Radio	Week-days	30	30	31	30	32	29	31	29	33	27
	Sun-days	22	21	23	21	23	21	22	21	23	22

(Source: NHK's National Survey on Listening and Viewing Rates)

TABLE 7 Average listening and viewing hours (people of 7 years or over)

Classification		Survey dates	a.m.	p.m.	Night time	per day
TV	Weekdays	Nov. 1975	48	48	2h 07	3h 43
		Nov. 1976	45	39	2h 05	3h 28
	Sundays	Nov. 1975	50	1h 20	2h 22	4h 32
		Nov. 1976	51	1h 07	2h 16	4h 14
Radio	Weekdays	Nov. 1975	16	13	9	38
		Nov. 1976	16	12	8	36
	Sundays	Nov. 1975	12	8	6	26
		Nov. 1976	12	9	7	28

(Source: NHK's National Survey on Listening and Viewing Rates)

CATV systems aimed at diversifying available programmes are found largely in areas adjacent to multiple channel regions.

Some CATV operators produce their own programmes, offering local information, and broadcast them side by side with the re-transmission of broadcasting programmes, but due to the large amounts of money needed for production such original programmes show little growth. As of the end of 1976, about forty facilities were presenting such programmes.

Most CATV systems, both for surmounting viewing difficulties and for programme diversification, are operated on the basis of dues paid by the subscribers; very few are operated for profit.

Chapter 4

NIPPON HOSO KYOKAI, THE JAPAN BROADCASTING CORPORATION (NHK)

1 OUTLINE

NHK, established in 1950, commenced operations by taking over the residuary assets of its precursor Nippon Hoso Kyokai. Since then it has relied for its financing solely upon the receivers' fees. NHK is fundamentally different from public corporations or government affiliated organizations which have become independent by separating themselves from the government or have come into being based on government starting funds.

The Broadcast Law has been very carefully formulated so as to ensure the autonomy and independence of the management of NHK, and to have the wishes of the people fully reflected in its operation. In particular, the management policy and other important matters related to NHK are decided by the Board of Governors, and all expenditures needed for the operation and maintenance of its businesses are to be financed by the receivers' fees collected by NHK itself. Also, the principle of freedom in broadcast programme compilation and rules for programme compilation are provided for in the Broadcast Law. At the same time, NHK is required by law to take steps to set up a body like the Broadcast Programme Consultative Council to give full expression to the wishes of the people.

NHK, which has as its ultimate objective contribution to the public welfare, finds its fundamental mission in the following rules, on the basis of which daily operations are conducted:

1 To broadcast excellent programmes in the field of news, education, culture and entertainment, so as to meet a large variety of people's wishes, and thereby

contribute to the elevation of the nation's cultural level;

2 To complete national networks in order that its broadcasting may be received all over the country, and at the same time, to establish broadcasting stations so as to provide regional and local services for the benefit of the inhabitants of local districts;

3 To conduct research and investigation necessary for the development of broadcasting and reception;

4 To co-operate with commercial broadcasters in supplying broadcast programmes and in the provision of technical assistance, in order to contribute to the advancement of the broadcasting world as a whole;

5 To operate overseas broadcasting services representative of Japan and to supply broadcast programmes to foreign broadcasters;

In addition, NHK sets up, on its own initiative, facilities for programme production and transmission necessary for broadcast operations. The microwave circuits required for network maintenance, however, are loaned from the Nippon Telegraph and Telephone Public Corporation.

2 STRUCTURE

(a) Board of Governors, Executive Organ, and Auditory Organ

In accordance with the provisions of the Broadcast Law, NHK has a Board of Governors as its highest decision-making organ, the President, Vice-President and Managing Directors as its executive organ, and Auditors as its Auditory organ, all of which come together to constitute the management system of NHK. The Board of Governors has the power and responsibility to decide management policy and other important matters relative to the operation of NHK businesses. The Board also has the highest rights in personnel management such as the appointment and dismissal of the President and Auditors, and the consent to the appointment and dismissal of the Vice-President and Managing Directors. Such a powerful organ has no parallel in other boards of this sort set up in Japanese corporations, and it is this structure that may be said to form the foundation ensuring NHK of autonomy and self-reliance.

(b) Organization of Board of Governors

The Board of Governors is composed of twelve Governors appointed by the Prime Minister with the consent of both Houses of the Diet (eight of them being appointed from each of eight districts, and the remaining four being appointed from throughout these districts). Consideration is given in their appointment to obtaining a fair representation in the fields of education, culture, science, industry and so on.

(c) Matters for Board decision

Important matters related to business management
* Budget of revenues and expenditures, operations and financial programme
* Settlement of accounts
* Plan for establishment of broadcasting stations, their inauguration, suspension, or abolition
* Standards for domestic broadcast programmes, standards for overseas programmes, and basic plans for programme compilation
* Alteration of articles of corporation
* Statutes of receiver's contract and standards for exemption of receivers' fees
* Issue of broadcast funds and obtaining funds on loan
* Rules governing business management
* Salaries, retirement allowances and social expenses of officers
* Other matters considered of special importance by the Board

Appointment of the President, etc.

* Appointment of President and Auditors
* Consent to the appointment of Vice President and Managing Directors
* Consent to nomination of members of Central Broadcast Programme Consultative Committee and Overseas Broadcast Programme Consultative Committee
* Dismissal of President and Auditors
* Consent to dismissal of Vice President and Managing Directors

The President represents NHK and presides over the business thereof as determined by the Board of Governors. The Vice-President (one) and Managing Directors (ten) assist the President in administering the operation of

NHK. In accordance with the provisions of the Articles of Corporation, several of the Managing Directors are nominated as General Managing Directors by the President, and also in accordance with the regulations concerning 'Director General of Engineering', one of the General Managing Directors is appointed Director General of Engineering. The President, Vice-President, Director General of Engineering, General Managing Directors and Managing Directors who compose the Board of Managing Directors as under the Broadcast Law, deliberate on matters requiring the decision or consent of the Board of Governors and matters deemed especially important by the President. The Vice-President and Managing Directors also deliberate on concrete measures relating to plans for future management, etc. at the Management Improvement Committee of NHK. Auditors report their findings to the Board of Governors. Also, in accordance with the Articles of Corporation, they may submit their views on matters needing improvement regarding the business of NHK to the Board of Governors or to the President.

(d) Structure of management

The management structure of NHK comprises the headquarters in Tokyo and various bureaux working as the general staff, General Broadcasting Administration, General Administration of Audience Services, Headquarters of Technical Administration and Construction, Staff Administration, ect., all working under the Board of Managing Directors. In addition, Japan is geographically divided into eight regional administrative districts. Excluding the region under the supervision of the Tokyo Headquarters, the remaining seven regions have their own regional headquarters in seven major cities, with the total number of stations reaching 62. Outside Japan, NHK has two general bureaux and 14 bureaux throughout the world (cf. the Organization Chart in Appendix).

Despite the expanding operations, attempts have been made to keep the number of employees to the minimum by achieving higher efficiency and labour saving devices through appropriate investments in mechanization. The total number of staff members has been kept down to 16,560 since the fiscal year of 1972.

3 FINANCE AND ADMINISTRATION

(a) System of budget compilation and account settlements

In accordance with the provisions of the Broadcast Law, NHK is required to obtain approval by the National Diet regarding its budget of revenues and expenditures, business project and financial programme. The approval of its budget by the Diet implies that the monthly amount of receivers' fees has been determined. The settlement of accounts for every business year also is submitted likewise for reporting purposes. Public control over the financial matters of NHK is primarily exercised by the Diet, representative of the people, and participation by the administrative organs in financial control is kept to a minimum. Such an arrangement forms the most salient feature of NHK, not to be seen in other Japanese public corporations.

The budget of revenues and expenditures, etc., for every fiscal year is independently prepared by NHK. It is decided on by the Board of Governors and then presented to the Minister of Posts and Telecommunications, who then submit it through the Cabinet to the Diet with his views attached. After deliberation by the Communications Committees of both Houses of the Diet, the Budget of Revenues and Expenditures, operations, and financial programme are approved en bloc. This approval is normally accompanied by supplementary resolutions of the two Diet Committees concerning their desires about the execution of the Budget.

In order to execute this budgetary plan, NHK formulates general rules and receives approval of them by the Diet in advance, as part of the Budget. Within this scope, NHK exercises autonomous control over the Budget, based on the decisions of the Board of Governors.

The settlement of accounts for every fiscal year, indicating the achievements of NHK's activities and its financial condition, together with the inventory, balance sheet, and profit and loss statement, all accompanied by a written explanation, are submitted to the Minister of Posts and Telecommunications after the decision of the Board of Governors. The Minister presents them to the Cabinet which will submit them to the Diet after the audit of the Board of Audit. As is the case with the budget of revenues and expenditures, the Communications Committees of both Houses of the Diet deliberate on the documents.

This arrangement for the Board of Audit, which is independent of the Cabinet, to audit the accounts of NHK seems to come from the consideration that these accounts should be subjected to public scrutiny since NHK is financed by the receivers' fees. The results of the audits are reported to the Diet.

(b) Financial management past and present

Since the fiscal year 1958 (April 1958 to March 1959), NHK has drawn up a series of long-range financial plans and continued its activities based on them. At present, NHK is putting into practice the three-year plan formulated in the year 1976. These long-term plans aim at the expansion of broadcast services for the benefit of the people, including the extension of NHK national networks, the compilation of richer programmes, and the stabilization of receivers' fees.

On the basis of the increased number of receivers' contracts for television, expansion centring around television broadcast services has been stimulated. On the other hand, modernization of operations, such as the introduction of automation and computerization by electronic devices, has been carried out side by side with efforts to economize operation costs and maintain the number of employees at a reasonable scale. As a result, NHK's financial conditions were comparatively stable before the year 1971. Thereafter, however, the number of colour television contracts began to show a very slow upward trend, resulting in considerably smaller increases of current revenue. To make the matter worse, commodity prices soared abnormally, causing current expenditures to surpass revenue. Beginning in 1972, current expenditure began to exceed ordinary revenue. In the fiscal years 1974 and 1975, the difference became much greater. This imbalance was partially offset by the appropriation of income from the sale of the Tokyo headquarters building in the year 1973, with the move to a new centre, and for the rest, by acquiring bank loans as business funds. By doing this it was possible to keep the amount of the receivers' fees at the same low level.

For the reasons just mentioned, the financial condition of NHK worsened. In the fiscal year 1975, the assets and capital of NHK showed a further decrease.

(c) Budget for the fiscal year 1976-7 and the three-year outlook

NHK was confronted with its most serious difficulties in the field of business management to date: the slowed tempo of increase in the number of receiver's contracts coupled with a phenomenal increase in expenditure due to recent economic fluctuations. To break the stalemate, in 1976, NHK took a step to revise the monthly amount of receiver's fees, based on its prospect three years ahead, from 1976 to 1978.

The proposed new rate was the colour contract fee, ¥710/month (up ¥245 compared with the previous year) and the ordinary contract fee, ¥420 (up ¥105). (In Okinawa, the rate was kept unchanged by special measures, with the colour contract fee at ¥400 and the ordinary contract fee at ¥250.) However, owing to the Lockheed scandal, Diet deliberations on the 1976 NHK budget, business project and financial programme was delayed. Consequently, this revision did not go into force until 1 June 1976, causing a considerable decrease in the revenue of NHK, and necessitating alteration to the three-year plan. The 1977 budget was formulated on the basis of the revised plan.

TABLE 8 Business Revenue and Expenditure, 1976-8. (in ¥100 million)

Division	Estimated 1976	Budget 1977	1978	Total 1976-8
Revenue including	1,920	2,108	2,171	6,199
receivers' fees	1,876	2,060	2,127	6,063
Expenditure	1,723	1,971	2,209	5,903
Difference between revenue and expenditure	197	137	- 38	296
Earmarked capital expenditure	110	117	69	296
Excess or deficiency	87	20	- 107	0

(d) Outline of business management

As previously mentioned, from 1958 on the business management of NHK was conducted on the basis of the annual business programme which was formulated in conformity with a series of long-range managerial prospects. In doing this,

NHK consistently aimed at the continued development of broadcast services provided for the nation, and the long-term stabilization of receivers' fees. At the same time, NHK, financed by the receivers' fees, gave high priority to efficient management and operation, taking various measures to achieve modernization and greater efficiency under its long-range plans.

To cope with expansion of television broadcasting services in particular, the introduction of automated machinery, unattended operation of equipment and facilities, and a computerized calculation system, was carried out, with the object of restricting the increase of expenses and personnel. Together with this, between 1963 and 1972, construction of the Broadcasting Centre was conducted in three stages. In 1973, when the Centre was completed, the former dual managerial functions which had been divided between the personnel stationed in the Tokyo headquarters building and those in the Broadcasting Centre were placed under unified managerial control. Meantime, in the organizational field, a series of innovations were carried out, aiming at the establishment of rationalized, functional and highly efficient systems to deal with expanding and diversified activities on a continuing basis. The centralized and simplified control system as seen today, with the Broadcasting Centre as its nucleus, thus came into being.

(e) Personnnel

From the initial five-year plan in 1958 and the growth of television broadcast services, the scale of NHK's operations expanded rapidly. Attempts were made, however, to limit the increase of personnel as far as possible by adopting efficient business operations, mechanization, appropriate staff arrangements, improvements in individual staff abilities, etc. The number of personnel reached a peak in 1972, totalling 16,560. There has been no increase on this figure since.

Total number of personnnel	
Staff members	16,210
Part time or temporary workers	350
Personnel structure	
Average age	39.0 years
Average years of service	15.5
Sex	Male 94.0%
	Female 6.0%

The 'Personnel System' systemizes job classifications along with promotions, salary increases, ability evaluation, and staff training. In particular, staff training covers a wide range of subjects and is designed to develop the abilities and widen the range of vision of each individual through group meetings and discussions.

4 NATIONWIDE DISTRIBUTION OF BROADCASTING SERVICES

NHK is actively engaged in developing and expanding both television and radio to achieve the mission of nationwide availability of broadcast services, incumbent upon NHK under the Broadcast Law.

(a) Television

(1) Areas with reception difficulties

With the start of regular television broadcasts in 1953, NHK formed its nucleus network linking major Japanese cities, and afterward continued to place emphasis on efforts to expand the network, aiming to eradicate the difficulty of reception in areas scattered throughout the nation.

As a result, the number of stations and other facilities now totals 2,495 GTV stations and 2,452 ETV stations operated by NHK, and 7,054 community reception facilities jointly established and operated by NHK and the viewers in the local areas with reception difficulties. With this, the number of households still having reception difficulties has been brought down to approximately 730,000.

In the construction and operation of these facilities, various rationalization measures have been taken through technical renovation and so on in order to control the trend of increased costs attendant on the smaller and more scattered areas with reception difficulties. In particular, beginning in 1975, NHK began to build simple, low-cost mini-relay stations, invented by NHK and with a minimum output of not more than 0.1 W. To minimize costs and meet the desires of local viewers, NHK has tried hard to construct these relay station facilities, antenna towers and so on, in co-operation with commercial broadcasters. Also, in the field of mini-relay stations, in order to expedite the establishment of commercial broadcasters' relay stations, NHK has made it a practice to allow commercial broadcasters free use of whatever equipment is capable of common utilization.

NHK intends to continue the effort to overcome reception difficulties using its present method of constructing relay stations and community reception facilities, and plans to reduce the number of households suffering difficult reception to about 400,000 by the end of the 1980 fiscal year. Thereafter, for those areas still with difficulties, there are hopes of utilizing the broadcast satellite now being developed, since these areas are so small in area and so sparsely scattered that it is considered very difficult, from both a financial and technical angle, to cover these areas by terrestrial networks.

(2) Reception difficulties in urban areas and their improvement

Television reception difficulties in urban areas in Japan have increased markedly in recent years, chiefly because of high-rise buildings. Households experiencing such difficulties number nearly 500,000 throughout the nation. An attempt to remove these difficulties is at present being made by, for example, setting up cable television broadcasting stations. NHK takes the basic stand that the constructors of the offending buildings should take responsibility for their obstructions and be urged to take voluntary action to solve the problem. At the same time, NHK provides technical guidance and carries out research, and is engaging in the development of technical devices to remove the difficulties. In parallel with this, NHK has been co-operating in the establishment and operation of such cable television broadcasting organizations as the recently established non-profit foundations which are now operating cable television broadcasts in Tokyo, Osaka and two other cities. In addition, since February 1976, NHK has been experimenting with television broadcast relay on the Super High Frequency Band.

Under the guidance of NHK or the Posts and Telecommunications Ministry, the idea of devolving the responsibility for reception difficulties upon those who have caused them has gradually established itself as a normal practice, although it is not yet provided for by law. Meanwhile, owing to such factors as the increased numbers of high-rise buildings, reception difficulties have been spreading and becoming increasingly complex, making the speedy and smooth solution therefore more and more difficult. Under the circumstances, the establishment of some effective countermeasures is becoming a problem which demands urgent solution, not only in the broadcasting circle but on a national level as well. In 1973, the Ministry of Posts and Telecommunications set up a research committee composed of men of learning and experience and with an assignment to study effective countermeasures for the reception difficulties in urban areas and remote regions. Two years later, the Committee submitted a report, based upon which the Ministry is now considering concrete policies for dealing with the matter.

(b) Radio

(1) Medium wave broadcasting

Ever since the start of medium wave broadcasts in 1925, NHK has been making continued efforts to expand its national networks. At present, the First Network comprises 173 broadcast stations, and the Second Network 141 stations. Covering 99.7 % and 99.2 % of the Japanese territory respectively, NHK's networks may now safely be said to be almost complete.

However, the expansion of its networks nearing completion has been attended by the growing interference of radio waves emanating from neighbouring nations. Night time reception of both NHK and commercial broadcasts has deteriorated year by year. NHK has taken countermeasures to improve the situation, increasing the output of four broadcast stations of the Second Network, including Tokyo, to 300 or 500 kW.

In November 1978 the AM frequency assignment programme which was adopted at the Administrative Radio Conference of the International Telecommunication Union in 1975 will take effect. Thereafter, it is believed that interference in medium wave broadcasts will be alleviated to a certain extent.

(2) FM broadcasting

Since the start of experimental FM broadcasts in 1957, NHK has steadily progressed in the establishment of a national FM network. Since 1969, when regular FM broadcasting began, NHK has striven for the completion of its national network, enabling each prefecture to receive independent broadcasts. The present NHK national network includes 464 FM stations, and reaches 96 % of households possessing receivers capable of receiving stereophonic broadcast. The percentage is higher in the case of households having sets whose capacity is limited to monophonic broadcast.

5 BROADCAST PROGRAMMES

(a) Domestic broadcasting

(1) Programme compilation and implementation systems

The Broadcast Law, while guaranteeing the freedom of broadcast programme compilation to all broadcasters and also clarifying basic standards of broadcast programmes, especially prescribes the following standards for NHK's broadcast programmes in addition to those common to both NHK and commercial broadcast programmes: 1, That the programmes be rich and good so as to satisfy diversified wishes of the people and to contribute to the elevation of the level of civilization. 2, That the programmes include local programmes in addition to national programmes. 3, That the programmes be such as to be of service in the preservation of past excellent civilization and the development and popularization of new civilization.

On the basis of these provisions of the Broadcast Law, NHK has established the Standards for Domestic Broadcast Programmes of the Japan Broadcasting Corporation, thereby clarifying the standards of domestic broadcast programme compilation. NHK also annually draws up Basic Plans for Domestic Broadcast Programme Compilation and follows these plans in actual programming.

Also, in order to maintain the appropriateness of its domestic broadcast programmes, NHK has a Central Broadcast Programme Consultative Committee and District Broadcast Programme Consultative Committees in eight regions covering the entire country. The members of these programme committees are nominated by the NHK President from among people of learning and experience. When the aforementioned Standards of Domestic Broadcast Programmes and the Basic Plans for Domestic Broadcast Programme Compilation are to be established or changed, the President must consult the Central Broadcast Programme Consultative Committee. Also, then basic plans for local broadcast programme compilation are to be established or changed, the advice of the District Broadcast Programme Consultative

Committees must be requested. Further, when these broadcast programme consultative committees deem it necessary for the maintenance of the appropriateness of broadcast programmes, they can state their opinions to the President. The Central Broadcast Programme Consultative Committee and District Broadcast Programme Consultative Committees, as a rule, meet once a month for such purposes. With regard to the composition of each broadcast programme consultative committee, moreover, constant infusing of new blood is promoted so as to reflect fresh views. In the conduct of the committees, too, deliberations are undertaken so as to absorb and reflect the views of various classes of people as much as possible. The activities of the broadcast programme consultative committees and summaries of their deliberation are made public as is seen fit.

In compiling domestic broadcast programmes NHK respects the views of the broadcast programme consultative committees, and for the purpose of accurately grasping the wishes of the people, conducts scientific listening polls at regular intervals, various other kinds of opinion polls and surveys on people's daily living timetable. NHK also uses, as reference data, views expressed at meetings of specialized committees on broadcast programmes and at study meetings, audience views presented to NHK directly in in writing (letters received in the fiscal year 1976 totalled 1,830,000) or by telephone, opinions presented to NHK's receivers' fees collectors on their collecting rounds, reports sent in by monitors specially appointed for the purpose, opinions and requests expressed at meetings between NHK officials and audiences which are held in various parts of the country, and reviews run in newspapers and magazines. In November 1976 NHK established Audience Advisory Councils in all the prefectures of the country, thus systematically and actively seeking the views of listeners and viewers.

Particularly in the compiling of school broadcast programmes and correspondence education programmes, analyses are made of the opinions expressed by those schools which are entrusted with the task of studying school broadcast programmes and also, the opinions expressed at the broadcast education study meetings which are held throughout the country. The basic plans of school broadcast programmes are referred to the local advisory committees on school broadcasting and to the central advisory committee.

In the implementation of broadcasting, the President serves as the responsible person representing NHK, and under him is posted the Director-General of Broadcasting who takes all responsibilities and has absolute power relative to the compilation of broadcast programmes. Thus the order of responsibility is clarified. Also, at headquarters is a Programme Inspection Bureau, set up as a voluntary programme inspection organ, under direct jurisdiction of the President. Similar inspection bureaux are also maintained at each regional headquarter in order to undertake inspection of all broadcast programmes in accordance with the provisions of the Broadcast Law and of the programme standards established by NHK.

The General Broadcasting Administration is in charge of compilation, production and transmission of broadcast programmes. Under the General Broadcasting Administration, the Broadcast Programme Control Centre, which discharges the essential functions of the Administration, uses electronic computer systems to carry out comprehensive planning of broadcast programme compilation, overall coordination of programme production and centralized control of programme transmission. In programme production, instead of the division system, excepting news programme production, the more flexible project team system is adopted for all categories. A similar system is also followed in production engineering. Meanwhile, specialized functions, such as announcing, camera work and design, are organized as separate groups in order to promote further repletion and training in these areas. Such measures are followed in order to cope with the increased volume work arising from the diversification of audience needs. At the same time, planned use of budgets, personnel, facilities and equipment is promoted so that, through clarification of the system of responsibility for programme compilation and improvement in work efficiency, those in charge of production may devote themselves to programme production as much as possible and display greater creative ability.

(2) Programme implementation

NHK now operates a General Television Service and an Educational Television Service in television broadcasting, and in Radio broadcasting, Radio First Network, a Radio Second Network and a FM service, altogether making for a total of five services. The basic characteristics of these five services are as follows. In the General Television Service, balanced programming covering news, education, culture and entertainment is undertaken and the programmes are presented for the public at large related to people's lifestyle. In the Educational Television Service, national broadcasts centred on systematic and continuing educational programmes are effectively compiled and cultural programmes are also presented. In Radio First Network, news, educational, cultural and entertainment programmes are compiled for general audience reception. In Radio Second Network, national programmes centred on educational items are compiled, which are designed for presentation to specific audiences. In the FM service, news and information programmes and programmes which give full play to the special characteristics of FM broadcasting are stressed.

The basic policy of NHK's network system can be said to be determined by the requests of the audiences who enjoy the benefits of broadcasting and sustain the foundation of the maintenance and operation of NHK. In that audiences, their time restricted by conditions of daily life, are likely to have diversified demands concerning broadcast programmes, and, in that such demands are likely to escalate in the future, the maintenance of the existing five networks is considered essential.

NHK's scheduled broadcasting hours include 17 hours and 30 minutes a day for the General Television Service (from 6.00 a.m. to 11.20 p.m. Mondays to Thursdays and Sundays, and from 6.00 a.m. to 12.00 midnight Fridays and Saturdays), 18 hours a day for the Educational Television Service (from 6.00 a.m. to 12.00 midnight), 19 hours a day for the Radio First Network (from 5.00 a.m. to 12.00 midnight), 18 hours and 30 minutes for Radio Second Network (from 5.30 a.m. to 12.00 midnight) and 18 hours a day for the FM Service (from 6.00 a.m. to 12.00 midnight). Incidentally, all programmes of the General Television Service are presented in colour, and all those of the Educational Television Service were scheduled to be presented in colour by the end of the fiscal year 1977. In the FM Service, a daily average of about 13 hours and 50 minutes are devoted to stereophonic programmes.

NHK's broadcasting of local programmes along with national programmes is an important task indispensable to a public broadcasting organization. In this connection, NHK presents local news and weather forecasts under the basic policy of prefectural unit broadcasting and also, information and cultural programmes closely connected with local conditions.

A national survey on broadcasting undertaken by NHK in May 1975 showed that NHK's television service ranked top among all forms of media available to the people. For instance, 81.6 per cent of people were found to be using NHK's television as a source of everyday news, 85.8 per cent cited it as a source of information on important functions and events, while 83.7 per cent depended on it for information on matters directly affecting them, such as typhoons and labour strikes. The people's reliance on NHK's news and information programmes and their assessment thereof are thus high. In order to meet such expectations on the part of the people, NHK, while providing a full service of news and information programmes, also undertakes quick reporting and priority reporting in impromptu programmes as the occasion demands. For this purpose, NHK maintains 174 correspondents' offices in Japan and sixteen overseas bureaux (two general bureaux and fourteen bureaux) which undertake mobile news coverage work, including satellite transmission of materials, as occasion demands. NHK does not stop at reporting only the phenomenal aspects of individual news items, but compiles various information programmes and commentaries that present impartial and comprehensive analyses of the backgrounds, the points at issue, and the effects and prospects of the various events. Some of these programmes are presented during evening prime time. News and information programmes scheduled in the General Television Service and the three radio services total about 17 hours and 40 minutes a day. In addition, flexible programming is undertaken as occasion demands, in accordance with the development of situations at home and abroad.

The role broadcasting is expected to play at times of emergency and disaster, such as earthquakes, tidal waves, storms and floods, is also important. Reporting of disasters is undertaken not only on the basis of various regulations, including the Fundamental Law of the Measures Against Calamity but also as a proper task of a public broadcaster. NHK overcomes unforeseen situations, such as power failures, flooding of the premises and blocked microwave channels, and manages to continue its operations - even relay broadcasting - through the use of private generators, and standby broadcasting equipment for emergency use.

On the basis of the Public Office Election Law. NHK also broadcasts the political views and personal histories of candidates standing for seats in the House of Representatives and the House of Councillors, and candidates for prefectural governorship. NHK further relays or broadcasts video recordings of public debates among the candidates, and also various other related programmes intended to deepen the voter's understanding of and interest in elections.

The attitude of attaching importance to educational and cultural programmes has been consistent throughout NHK's history. Based on years of experience and accumulated knowledge, NHK, from its position of guaranteed impartiality and equity, makes good use of its nationwide networks to effectively contribute towards the correction of regional gaps in education and towards the availability of equal educational opportunities and life-time education.

Educational programmes can be classified roughly into school broadcast programmes (112 programmes presented for 33 hours and 55 minutes a week on the Educational Television Service and 27 programmes presented for 25 hours and 50 minutes on Radio Second Network), correspondence education programmes (24 items presented for 12 hours on the Educational Television Service and 66 items presented for 22 hours on Radio Second Network) and social education programmes. According to a survey taken on the utilization of school broadcasting, the distribution of receiving sets and the rate of their systematic utilization are quite noteworthy. In particular, the use of colour television sets and the school broadcasting services at educational institutions for the handicapped are increasingly expanding.

The cultural programmes presented probe deeply from various angles into subject matters closely related to social life, meeting diversifying audience needs. In addition, campaign and related programmes that deal with resources, energy, commodity prices, environment and social welfare are flexibly compiled in concentrated form.

NHK's educational and cultural programmes enjoy a high reputation at overseas contests, such as the Asia-Pacific Broadcasting Union Prize and Italian Prize contests, and there are many requests from overseas countries for distribution of such programmes.

In the field of entertainment programmes, the development of wholesome programmes which the whole family can enjoy is promoted, and endeavours are made to provide entertainment programmes that introduce traditional entertainment arts, such as traditional Japanese music and

Bunraku puppet plays, and that help create new culture.

NHK also procures good overseas broadcasting programmes through purchases or exchange and presents them actively and concentratedly to meet the wishes of the people.

(3) Incidental activities

NHK undertakes various incidental operations for the sake of discharging its mission more smoothly and speedily. One such operation is the promotion of organizations needed for fulfilling and improving broadcasting programmes and for smooth broadcasting operations. For example, NHK subsidizes the NHK Symphony Orchestra Foundation, and, by ensuring broadcast performances and helping the orchestra to raise its technical level of performance, provides for a richer content of broadcasting programme. Also, for the purpose of providing correspondence education programmes and of promoting their use, NHK established the NHK Gakuen Correspondence High School in 1962 and has continued to support it since with annual subsidies.

NHK also sponsors lecture meetings and concerts by noted men of culture and musicians, both from Japan and overseas countries, largely for the purpose of their later broadcasting. Especially since the completion of the NHK Hall in 1973, NHK has invited such world famous groups as the Vienna Philharmonic Orchestra, the Berlin Philharmonic Orchestra and the Italian opera, and their performances have been highly successful.

NHK endeavours to promote the use of educational and cultural programmes through active sponsorship of various study meetings and training sessions, in tie-up with broadcast education study and other educational organizations, and also with administrative agencies related to education. As a result, the number of television and radio receivers used in schools and the rate of utilization of school broadcasting programmes are, as stated above, annually increasing.

Also, in order to boost the use of educational and cultural programmes BHK, in conjunction with local administrative agencies and other pertinent organizations, promotes campaigns for making use of broadcasting in everyday life. In addition to thus endeavouring to train leaders and publicize programmes, NHK publishes various textbooks intended to facilitate the use of broadcast programmes. NHK's publication programme in the 1976 fiscal year included 76 items totalling 39,800,000 copies.

NHK also presents audience participation programmes from the NHK Hall and from various parts of the country in order to deepen the interest of the audiences in broadcast programmes as a whole.

(4) Miscellaneous

Since the inauguration of commercial broadcasting in Japan, NHK has been providing co-operation for the commercial broadcasters so as to improve broadcasting as a whole. The significance of such co-operation was defined in a revision of the Broadcast Law in 1959, and thus NHK provides the co-operation on the basis of legal provisions.

NHK supplies the commercial broadcasters with what is needed for compiling and broadcasting programmes. With the permission of the Minister of Posts and Telecommunications NHK contributed part of its assets when the Broadcast Programming Centre was established in 1968 for the purpose of promoting enrichment and improvement of the educational and cultural programmes of the commercial broadcasters. Since then NHK has been making monetary contributions to the Centre. NHK also supplies the Centre with educational and cultural programmes. These programmes are lent out to individual commercial broadcasters by the Centre and are contributing to the improvement of broadcast programmes.

In the field of engineering, NHK undertakes research and investigation entrusted to it by commercial broadcasters and makers, provides technical assistance for commercial broadcasters and also training for commercial broadcasting personnel.

When NHK sold the land and buildings of the Tokyo headquarters in 1973, in order that the funds may be put to the most effective use, it decided to allot ¥12,000 million of the proceeds to the establishment of the Hoso Bunka Foundation (Broadcasting Culture Foundation), and having obtained the authorization of the Minister of Posts and Telecommunications, completed establishment in the following year, 1974. The Foundation promotes the development and improvement of broadcasting culture, thereby contributing to the advancement of broadcasting and reception by subsidizing and assisting such activities as technical research and development, international co-operation, research and investigation in legal, economic, social and cultural spheres, promotion and preservation of culture.

In the implementation of everyday services, too, NHK co-operates with the commercial broadcasters. For instance, NHK builds certain television relay stations jointly with commercial broadcasters, undertakes joint coverage work with commercial broadcasters in reporting big events like the Olympics, works jointly with commercial broadcasters based in Tokyo in the reception of satellite transmissions and helps maintain the Broadcasters' Council for Better Programming jointly established with the National Association of Commercial Broadcasters in Japan.

(b) Overseas broadcasting

As stipulated by the Broadcast Law NHK now operates overseas broadcasting. The government annually issues an overseas broadcasting directive specifying broadcasting matters and coverage areas. The government bears the cost required for government-specified operations from the national budget.

The Broadcast Law prescribes that in compiling overseas broadcast programmes NHK 'shall exert its effort in contributing to the promotion of international friendship and to the development of economic interchange with foreign countries, by cultivating and popularizing the correct recognition of our country by introducing our civilization, industry and other state of things; at the same time, it shall endeavour to give appropriate entertainments to our nationals in foreign countries' (Article 44-5, Broadcast Law).

As in the case of domestic broadcasting NHK has established the Standards of International Broadcast Programmes of the Japan Broadcasting Corporation in order to clarify the standards of overseas programme compilation. NHK sets up basic plans for overseas broadcast programme compilation each year and endeavours to enrich and improve its broadcast programmes.

At present NHK's overseas broadcasting service, Radio Japan, presents programmes for a total of 37 hours a day, broadcasting in 21 different languages. The overall service is divided into subservices including the Regional Service and the General Service.

The Regional Service is directed towards different regions of the world. The programmes for this service are compiled taking into consideration the political system, ethnic composition, religion and customs and manners of specific regions. These are presented mainly in the local language, with consideration given to good radiowave propagation time and the suitability of the times of the day in the living habits of the local people.

The General Service was inaugurated in 1960 with the target of achieving an all-day quick-reporting system. Programmes of this service, largely consisting of news and news commentaries, are presented every hour on the hour for 15 minutes and three times a day for 30 minutes in English and Japanese.

Signals of NHK's overseas broadcasting are directed throughout the world from the Yamata Transmitting Station of the Kokusai Denshin Denwa Company (International Public Telecommunications).

While contributing to enrichment of domestic programmes, NHK also broadcasts programmes introducing Japan's social conditions, arts and so forth through broadcasting organizations of overseas countries. Also, in order to promote cultural interchange, NHK exchanges broadcast programmes in films and video or sound recordings with foreign broadcasting organizations, at the same time actively participating in programme exchange activities of the Asia-Pacific Broadcasting Union (ABU) and the European Broadcasting Union (EBU). International programme exchanges via satellite transmissions are also undertaken on appropriate occasions.

In addition, NHK supplies broadcast programmes to NHK International, a division of NHK which is engaged in international distribution of NHK programmes and upkeep of the NHK international programme library, and thereby works towards the enhancement of international understanding.

NHK has concluded agreements related to programme exchange and the like with fourteen overseas broadcasting organizations, with a view to promoting closer co-operation in this area.

NHK also sponsors the Japan Prize International

Educational Programme Contest every other year as a means of contributing to the enhancing of the world's educational programmes and promoting international co-operation and understanding.

6 RESEARCH AND INVESTIGATION

(a) General situation

The Broadcast Law expressly stipulates that NHK undertake research and investigation necessary for improvement and development of broadcasting and reception thereof. In so doing, it respects the opinions of outsiders related to the research and investigation thus constantly providing for the maintenance of high levels in broadcasting culture and broadcasting engineering in a period of rapid development. The law also provides that the results of such research and investigation be utilized in the interest of the general public as far as possible.

NHK maintains the Radio and Television Culture Research Institute and the Public Opinion Research Institute for broadcasting culture research and investigation, and the Technical Research Laboratories and the Broadcasting Science Research Laboratories for broadcasting technology research and development. These institutions undertake research and investigation using as reference the opinions of research committees composed of members who do not belong to NHK.

In the budget of the fiscal year 1977 NHK designated ¥6,000 million as a research and investigation fund, including personnel costs and depreciation. This amount accounts for 3 per cent of NHK's working expenditure. About 500 researchers are engaged in technical research.

(b) Broadcasting culture research and investigation

The Broadcast Law requires NHK to conduct scientific listening polls at regular intervals, so as to reflect the wishes and opinions of the people at all times, and to make the results of such polls public. And NHK endeavours to reflect the results of such surveys in programme compilation and production. In the nation wide Time Budget Survey, the subject of investigation is how people spend the 24 hours of the day on weekdays, Saturdays and Sundays. It is thus a large-scale survey aimed at grasping details and changing aspects of the life of the Japanese people. Since the first survey in 1960, similar surveys have been undertaken in 1965, 1970, 1973 and 1975, and the results of these surveys are considered highly important, not only by those involved in broadcasting.

Also, as research and investigation contributory to the enrichment of programme contents, NHK undertake studies related to programme planning and presentation, and investigation of broadcasting situations both in Japan and overseas countries. In addition, it undertakes study aimed at a theoretical systematization of broadcasting, manages a broadcasting culture library, set up for systematic collection and preservation of various cultural properties in sound and video recordings, and also, has recently compiled a history of fifty years of broadcasting in Japan.

The results of researches, in conformity with the intent of the Broadcast Law, are made public through announcements in the press media, publication in magazines and distribution among interested circles.

(c) Broadcasting technology research and development

In 1930, shortly after inauguration of broadcasting in Japan, NHK established the Technical Research Laboratories. Since then, the Laboratories have been conducting high-level research that has remained at the forefront of Japanese broadcasting technology. Work undertaken at the Technical Research Laboratories covers all sectors of broadcast engineering, ranging from basic theories to practical applications.

Divided into various areas, NHK's research covers, first, improvement of existing broadcast engineering, such as colour television engineering, acoustic engineering, reception improvement engineering, etc.; second, development of new broadcasting systems, such as satellite broadcasting, still picture broadcasting, four-channel stereophonic broadcasting, etc.; third, new techniques that sustain broadcasting, such as super high-frequency, recording and data processing techniques; and finally, basic work closely related to broadcast engineering, such as audio visual science and solid state properties. Outstanding results of researches in recent years include development of a compact and high-performance colour iconoscope 'Saticon' and of a high-sensitivity but low-cost receiver for satellite broadcasting. As in the case of broadcasting, culture research and investigation, the results of technical research is made public through reports at academic meetings, publication in magazines and distribution among interested circles.

NHK also provides co-operation for commercial broadcasters, and other outside organizations like local autonomous bodies in such forms as the designing of broadcasting equipment and guidance in the manufacture thereof. In addition, NHK authorizes use of 676 patents and utility model registrations (including 127 foreign patents) it holds, thus promoting wide utilization of the results of its researches by this means as well.

7 AUDIENCE RELATIONS

(a) Receivers' fee system

The financial basis for NHK's business operations is the receiver's fee system, under which the audience, that is, the people, bears the costs directly.

The receiver's fee system is based on the provision of the Broadcast Law (Article 32 Section 1) which obligates 'any person who is equipped with a receiving equipment capable of receiving the broadcasting provided by the Corporation (NHK)' to conclude a contract with NHK with regard to the reception of its broadcasting, but unlike the prewar receiving equipment licensing system or the reception licensing systems of foreign countries, no punitive or other legal disciplinary clauses are provided.

The existing receiver's contract system was established in 1968 and is divided into two classes, one being contract (ordinary contract) that does not include reception of colour television and the other, contract (colour contract) that includes reception of colour television programmes. In that year, the Broadcast Law was amended to nullify the receiver's contract related to radio reception only.

The monthly amount of receiver's fee is set every business year when NHK's budget of revenues and expenditures is approved. The current monthly amount for the ordinary contract is ¥420 and that for the colour contract is ¥710 (special rates are applied in Okinawa prefecture).

The provisions of the contract are set forth by approval of the Minister of Posts and Telecommunications. Under the existing provisions, as a rule, the contract is concluded by the household. With regard to receiving equipment installed in business premises, each room or each car is considered as the relevant unit.

Receivers' contracts are concluded through NHK's own collectors, or commissioned collectors, or post offices entrusted with the task, or commissioned canvassers.

Receivers' contracts (paying contracts) total 26,430,000 of which colour contracts account for 23,120,000 while ordinary contracts for 330,000 (as of the end of fiscal 1976).

Growth in the number of receivers' contracts is gradually slowing down. This is considered to be due to approaching saturation.

Receivers' fees are collected by NHK's own collectors, commissioned collectors and post offices entrusted with the task, but in order to cope with an average population mobility rate as high as 12 per cent for the country, and with diversification of social environment and living modes, utilization of the bank account transfer payment and advance payment systems are being encouraged for the sake of improving the efficiency of fee collection. The 1976 collection rate was 96.5 per cent.

Exemption from the receivers' fee is based on the standards set by approval of the Minister of Posts and Telecommunications. From the standpoint of social welfare and educational policy, social welfare facilities, educational institutions and needy households are exempt from the entire amount, while those with audio-visual impediments or serious physical handicap are required to pay half the amount only.

(b) Interchange with audiences

Mutual understanding and trust between audiences and NHK form the basis of existence of the public broadcasting organization. For this reason, NHK's business management and broadcast programme contents are publicized widely, while at the same time, various means are employed for soliciting the views of the audiences and reflecting them in business management.

Public Relations Programmes: Programmes intended to provide answers to the people's opinions and requests and various other public relations programmes are presented. One such programme is titled 'Audience and NHK', in which responsible staff members of NHK exchange ideas and views directly with the audience.

Audience Advisory Council: For the purpose of more strongly reflecting the people's opinion on its business management, NHK has been holding meetings with audiences in various parts of the country totalling as many as 1,800 a year. NHK also has been holding meetings with well-informed persons to hear their views on NHK. In addition, in 1976, NHK newly inaugurated Audience Advisory Councils at 53 places throughout the country, for the purpose of soliciting views on a nation wide scale and deepening mutual understanding.

Each Audience Conference is attended by about twenty local inhabitants, selected so as to represent various walks of life. These people express their views on NHK's business programme, budget and so forth. Although the attempt is new, the meetings are making a good start.

Audience Relations Centre: NHK maintains Audience Relations Centres at the Broadcasting Centre in Tokyo and each regional headquarter. The aim of the centre is to process inquiries and so forth from the audiences and to absorb frank opinions. The Broadcasting Centre in Tokyo has a visitors' course designed, from the beginning, for the convenience of people who come to look over the facility. This course is open to the public practically all the year round. Visitors who come to look over the Broadcasting Centre and the adjoining NHK Hall total about 3 million a year. All NHK broadcasting stations in the country also open their buildings to visitors who, as far as possible, are invited even into studios.

Public Relations Publications: Various kinds of pamphlets are issued and distributed for publicizing NHK's business operations. Public relations are also undertaken through advertisements in newspapers and magazines. The publications include English pamphlets titled 'This is NHK' and 'Educational Broadcast of NHK'.

8 INTERNATIONAL RELATIONS

(a) Participation in regional broadcasting unions

NHK maintains a close relationship with broadcasting unions of various regions and co-operates in internationa- activities, such as programme exchange, as a means of contributing to the development of world wide broadcasting circles. NHK is now a full member of the ABU and an associate member of the EBU.

NHK has co-operated in the establishment of the ABU since 1957, and after the organization was officially inaugurated in 1964, NHK has been actively co-operating in the exchange of programmes and materials, information exchanges and management of the provisional co-ordinating centre related to satellite broadcasting. NHK was admitted as an associate member of the EBU in 1951, and while dispatching representatives to the general assembly, various committee meetings and working group meetings incidental to the committee sessions, it also participates actively in programme exchange.

(b) Co-operation with foreign broadcasting organizations

In addition to co-operating in the various activities of the ABU and the EBU, mentioned above, NHK also co-operates with various foreign broadcasting organizations on a bilateral basis. Excepting cases of government involvement, this type of co-operation is carried on directly between the parties concerned, each helping the other in programme production and procurement of materials.

Depending on the country or the broadcasting organization, however, permanent agreement on co-operation is officially concluded. The substance of such agreement generally includes exchange of news or other programmes, co-operation in coverage work and exchange of materials. As of the end of March 1977 NHK maintained such agreements with fourteen broadcasting organizations.

(c) International exchange of programmes

NHK exchanges programmes in film, video tape or sound recording form with foreign broadcasting organizations so as to enrich its domestic programmes and to introduce Japan's social conditions and arts through foreign broadcasting organizations, thereby promoting cultural interchange. NHK also implements, from time to time, international exchange of programmes via satellite transmissions.

Aside from the foregoing, NHK supplies the NHK International (established in September 1967), which is engaged in international distribution of NHK's programmes and management of NHK's international broadcast programmes, with broadcast programmes so as to contribute towards better international understanding.

The Japan-US Television Programme Festival has been held alternately in Japan and the USA since 1973 for the purpose of promoting programme exchange between the two countries, thereby helping to deepen mutual understanding between the two peoples. NHK exhibits its programmes at this festival (from the fourth meeting named the Japan-US Television Executives' Conference).

NHK also inaugurated the Japan Prize International Educational Programme Contest in 1965 in commemoration of the fortieth anniversary of broadcasting in Japan and of the holding of the Second International Conference of Broadcasting Organizations on Sound and Television School Broadcasting in Tokyo. The contest has been held once a year up to 1975. This is the world's first and only international contest in educational programmes. Open to participation by the broadcasting organization of the countries and regions participating in the International Telecommunication Union (ITU), it contributes to the improvement of world educational programmes and to increased international understanding and co-operation. Also, in order to assist in the improvement and development of educational programmes through circulation of the achievements of this contest among the world's broadcasting and educational circles, in 1974, the Japan Prize Circulating Library was established. Under this system, excellent programmes selected at the contest are sent to regional centres throughout the world and circulated for use in training and study.

(d) Overseas technical co-operation

NHK actively provides technical co-operation for developing countries, principally on a government-to-government co-operation basis. In this connection, NHK dispatches its broadcasting and engineering leaders to overseas countries for providing guidance in the construction of broadcasting centres, planning of broadcasting network expansion, training of personnel and in programme production.

NHK also provides training at its Central Training Institute for overseas broadcasting personnel in educational programme production and in television and radio broadcast engineering, all for the purpose of contributing to interchange with these countries.

In addition, NHK offers technical guidance in broadcasting facility designing and so forth for visitors from overseas countries. Furthermore, such achievements of NHK's broadcast engineering researches as have been registered as patents or utility models are licensed to requesting overseas organizations under reasonable terms.

Chapter 5

COMMERCIAL BROADCASTERS

1 ESTABLISHMENT AND GROWTH OF COMMERCIAL BROADCASTING SERVICES

Commercial broadcasting services were authorized in Japan in June 1950, when the Radio Law and the Broadcast Law were enacted. These laws were enacted for the purpose of releasing radiowaves, which had been monopolized by the Government, for civilian use.

As a result, NHK's monopoly on broadcasting came to an end and an era of parallel operations of broadcasting by NHK and commercial broadcasters was ushered in.

In April 1951 provisional licences were granted to sixteen commercial radio companies for the first time. These companies inaugurated broadcasting operations one after the other, following in the wake of the Chubu Nippon Broadcasting Company (CBC) in Nagoya and the Shin Nippon Broadcasting Company (NJB) in Osaka (now Mainichi Broadcasting System - MBS), which pioneered the field in September of the same year.

As radio broadcasting has already been operated by NHK over its national networks, receiving sets were in use throughout the country in adequate numbers. This allowed the commercial broadcasters to make a smooth start.

Commercial television service was inaugurated about two years later, in August 1952, by Nippon Television Network Corporation (NTV).

Unlike the case of radio, commercial television started at about the same time as NHK. The greatest concern in this area, therefore, was distribution of receiving sets.

In those days, television set production was still inadequate and the sets available were costly. Furthermore, Japan's economy had not yet recovered from the postwar confusion and national life continued to be in the doldrums.

For several years after commencement of NTV's service television remained accessible only to the wealthier classes of people, or to those people who would gather in the plazas or restaurants and the like to view receivers set up there. General distribution amongst homes was very slow.

None the less, in 1957, the Ministry of Posts and Telecommunications took a step towards nationwide operation of television and licensed thirty four commercial companies all at once. This laid the basis for establishing at least one television station in each prefecture and two to four stations in each major city. As a result, television took its first stride towards rapid expansion.

With the establishment of the system of parallel operation of television service, NHK and commercial broadcasters, the receivers' fee system was left to NHK for operating its service while the commercial broadcasters, for their finances, were left wholly dependent on income from advertisements.

In general, mass communications media are sustained in some degree by income from advertisements. In the case of commercial broadcasting, however, 100 per cent of income is dependent on advertisements since service is provided in a blanket coverage form. Initial anxiety about the success of commercial broadcasting proved groundless and commercial broadcasting launched out on a course of amazing development.

The development can be considered attributable largely to the fact that the economic growth of Japan began at about the same time, and to restraint of competition from new enterprises because of the need for obtaining government's licence. Not to be overlooked, however, were the efforts made in tapping the value of the media as a profit making enterprise.

As of the end of March 1977 Japan's commercial radio and television broadcasting companies numbered 107, and all of them, excepting one, are incorporated profit-seeking organizations. The one exception is the radio company (Kyokuto Hoso) organized as a foundation and established in Okinawa when the islands reverted to Japan in 1972. In the twenty five years of commercial broadcasting, two non-profit-seeking foundations other than Kyokuto Hoso have existed, but, unable to meet the costs, are now reorganized as profit-seeking entities.

The commercial broadcasters comprise 53 radio broadcasting companies and 90 television companies, but as 36 companies operate both radio and television services, radio-only companies total 17 and television-only companies 54, totally 107 in all.

Commercial broadcasting went through a number of changes before the present form of broadcasting was arrived at. First of all, radio stations were set up in major cities and extended into other cities from time to time. Five years later, by around 1955, the basic plan of establishing one station in every prefecture and a plural number of stations in the major cities had been accomplished. Subsequently, one new station each was established in Sapporo, Nagoya and Osaka in order to cope with increased demand for radio in the major cities.

At the same time, prefectural unit broadcasting stations were established in five prefectures in the Kanto (eastern Japan around Tokyo), Tokai (central Japan around Nagoya) and Kinki (mid-western Japan around Osaka and Kyoto) regions, which were high-power, wide-coverage broadcasting areas. In 1971 there was a broad increase in power output as a measure against interference by foreign signals, but, except for low-power re-broadcasting stations, no new radio stations have been established since 1963. This is the result not so much of the lack of demand for radio, but of a worldwide shortage of medium waves.

Other measures related to radio included licensing, in 1969 and 1970, of four commercial FM stations in Tokyo, Nagoya, Osaka and Fukuoka to meet public demand for commercial FM stations. Nationwide establishment of similar stations is still in abeyance. This stems from the government's policy of, in the future, substituting FM broadcasting for medium wave broadcasting, a policy based on the possibility that increased interference by foreign signals might eventually prevent adequate functioning of radio broadcasting.

Television was inaugurated in 1953. In 1957, the

government came out with the policy of establishing one station in every prefecture and two to four stations in major cities. Full-scale penetration of television, therefore, can be said to have been achieved after commercial television stations were established in all parts of the country.

Television stations established in various parts of the country under mass licensing were mostly managed by radio broadcasting companies already in operation. In the background to this was the obvious advantage of making use of the experience, equipment and profit-making ability of the radio companies and also, their desire to protect the future of the radio service.

In 1959, the Nippon Education Television Company (NET - but now TV Asahi) and the Fuji Television Company opened their stations in Tokyo, bringing the number of commercial television stations to four. As all the provincial cities had only one station - except Osaka with four, Fukuoka with three and Sapporo and Nagoya with two each - competition between key stations in securing network participating stations was intense.

The next change in the policy of television station establishment came in 1967: the idea of setting up two commercial stations in every prefecture. As a result, mass licensing of UHF stations was undertaken in several stages.

This was followed by the establishment of one prefectural unit station in each prefecture, even in the wide-coverage broadcasting areas, for the purpose of promoting closer affinity of programmes with the local people. These policies successively led to a rapid increase of commercial television stations which has continued up to the present day.

Meanwhile, exclusively educational stations which, for financial reasons, became incapable of fulfilling a substantial role, were discontinued and all the commercial stations became stations presenting general programmes.

As an index of the progress of commercial broadcasting services, data on the increase in income from television and radio advertisements is cited below.

2 COMMERCIAL BROADCASTING SERVICES AND LEGAL DISCIPLINE

As the so-called business licence system, which directly licences the broadcasting business itself, is not followed in Japan with respect to broadcasting, position as a commercial broadcaster is acquired by obtaining a licence for a broadcasting station as a type of radio communication station.

In that the social impact of broadcasting is extremely strong, licensing of broadcasting stations, unlike the case of ordinary radio stations, is done after screening as to whether the applicant is capable of meeting the provisions set forth in the Fundamental Standards for Establishment of Broadcasting Stations (Posts and Telecommunications Ministry Ordinance). Under these standards, careful examination is made of the applicant's equipment, his financial ability to maintain the broadcasting service and the adequacy of the broadcasting service from the viewpoint of public welfare.

As a result, the commercial broadcaster is obliged to meet, in addition to regulations set forth in the Radio Law for general radio stations, special qualifications as a broadcaster and a number of requirements related to the performance of the broadcasting service. Principal regulations are:

1 Any person who undertakes a broadcasting service shall not have alien affiliations either with respect to personnel or to capital (exclusion of alien affiliations).

2 The officers and investors of the business entity shall as far as possible have ties with the community in which the business entity operates (close local ties).

3 Excepting the case of concurrent radio and television service operation and of operating relay stations, no broadcaster shall control more than two broadcasting stations (mass communication concentration foreclosure).

4 No simultaneous management or control of radio, television and newspaper business shall be permitted (mass communication concentration foreclosure).

5 The effective term of licence shall be three years, and in order to continue with the broadcasting service,

application for licence renewal shall be submitted every three years at the specified time.

6 While freedom of broadcast programme compilation is guaranteed, the following working rules shall be observed on a voluntary basis:

Broadcasting

* Shall not disturb the public security and good morals and manners;
* Shall be politically impartial;
* Shall broadcast news without distorting facts;
* As regards controversial issues, shall clarify the point of issue from all angles possible;
* Excepting broadcasting stations licensed to broadcast special programmes, the educational, cultural, news and information and entertainment programmes must be presented in a balanced form (general television, 10 per cent or more educational, 20 per cent or more cultural programmes, TV Asahi and Tokyo Channel 12, 20 per cent or more educational and 30 per cent or more cultural programmes, and for FM service, 70 per cent or more of stereophonic programmes are set as conditions for licensing).

7 Standards for broadcast programme compilation according to the type of programme and the audience shall be adopted and programmes be compiled and presented in accordance with the standards.

8 To strive for the appropriateness of broadcast programmes, commercial broadcasting companies shall maintain broadcast programme committees.

9 No agreements related to supplying broadcast programmes, which contain clauses leading to receipt of supplies from a specific supplier only, shall be concluded.

The nature of the above regulations is in part explained by the expectation that commercial broadcasting would be useful in fostering democracy, and that the authorities were unable to anticipate the future state of commercial broadcasting at the time of the law's enactment.

As commercial broadcasting developed to a level comparable to NHK and its impact on the public has increased, amendments to the law so as to match the broadcasting system with the existing state of affairs have frequently been passed.

At present, administrative guidance is provided in a number of areas for supplementing the inadequacy of the existing broadcasting system. The more important of these include 1, construction of relay stations for solving the question of reception difficulty, 2, reinforcement of voluntary control of broadcast programmes, and 3, expansion and enrichment of local programmes.

3 STATIONS

The establishment of individual broadcasting stations is based on the channel plans indicated in the administrative policy of the Ministry of Posts and Telecommunications.

These channel plans are intended to provide broadcasting services that will meet the people's demand for quantity and quality in broadcasting as far as is technologically possible. The plans are drafted separately for medium wave broadcasting, television broadcasting and FM broadcasting. The establishment of commercial broadcasting stations as laid out in these plans is undertaken as described below.

(a) Medium wave broadcasting

Under the channel plan for medium wave broadcasting, nationwide networks of NHK's Radio First and Radio Second services are allowed. With respect to commercial broadcasting, stations are to be established in the order of cities with the larger number of households, so that reception will become possible in all the important areas of the country. Also, comparative assessment is made from the standpoint of administrative, economic and cultural importance, and station establishment is to be authorized under special consideration in areas deemed as requiring such stations.

Under this policy, commercial medium wave broadcasting stations (key stations) have been established as follows:

Tokyo, Osaka, Okinawa:	3 stations each
Sapporo, Nagoya, Fukuoka:	2 stations each
Other areas:	1 station in the central city of each prefecture except for the Gunma, Saitama, Chiba, Mie, Shimane and Saga prefectures.

The output of these stations ranges as follows:

Tokyo stations:	100 kW
Osaka, Sapporo, Nagoya and Fukuoka stations:	50 kW
Yokohama station:	30 kW
Sendai, Kyoto, Kobe and Hiroshima stations:	20 kW
Other stations:	5 kW

There are a number of medium wave relay stations to supplement the service provided by the key stations. Including these, therefore, medium wave stations of commercial broadcasters total 177.

However, as high-power, medium wave stations are to be found in increasingly large numbers in countries neighbouring Japan, powerful signals beamed from such stations invade the sky over Japan at nights, thus markedly impeding the night-time service of Japan's medium wave stations.

(b) Television

While channel plans for television have undergone various changes along with the diffusion and development of television, the following four points constitute the basic concept:

1 In frequency allocation, consideration is given so that the specific frequency allocated will be receivable by as many viewers as possible in as wide an area as possible.

2 Consideration is given so that NHK's General TV and Educational TV services and the commercial services may be enjoyed throughout the country under approximately equal conditions.

3 Priority allocation is made to areas with higher population density and allocations to other areas are made in an order of descending priority.

4 In areas where regional unity with respect to political, economic and cultural conditions and people's lifestyle is found, measures are to be taken so that the inhabitants will enjoy the same broadcasts under approximately equal conditions.

In line with this basic policy, from 1953 on, commercial television stations were established successively in the leading cities of the country, and ten years later, in 1963, a nationwide commercial broadcasting network was completed.

Because the establishment of commercial stations was undertaken with priority given to the economic strength of the areas, however, while commercial broadcast programmes of two to four stations became available in economically powerful, large cities and their peripheral areas, in other areas, no more than one commercial programme was available.

As commercial broadcasting is a business enterprise operated independently in respective regions, the question of the service area of each enterprise becomes important. The service areas are determined as follows under the aforementioned basic policy:

1 Stations in the Tokyo-Yokohama Wide Coverage Sphere:
   Tokyo Metropolitan area, and Kanagawa, Chiba Gunma, Tochigi, Ibaraki and Saitama prefectures.
2 Stations in the Chukyo Wide Coverage Sphere:
   Aichi, Gifu and Mie prefectures.
3 Stations in the Kyoto-Osaka-Kobe Wide Coverage Sphere:
   Osaka, Kyoto, Hyogo, Nara, Wakayama and Shiga prefectures.
4 Stations in Tottori and Shimane prefectures:
   The area covering both prefectures.
5 Stations in other regions:
   The total area of the prefecture in which a station is located.

The number of commercial broadcasters in various regions is as given below:

1 Principal regions (four or more under the channel plan)
   Tokyo: 5
   Sapporo, Sendai, Nagoya, Osaka, Hiroshima, Fukuoka: 4 each
2 Semi-principal regions (three under the channel plan)
   Shizuoka, Niigata, Nagano: 2 (in all three areas, the third station is under preparation)
3 Tottori and Shimane prefectures: (three for the two prefectures under the channel plan)
   Tottori: 1
   Matsue: 2
4 The prefectures within the wide-coverage spheres

(One for every prefecture under the channel plan)
Maebashi, Chiba, Yokohama, Gifu, Tsu, Otsu, Kyoto, Nara, Wakayama: 1 each
(None in Mito, Utsunomiya and Urawa)

5 Other prefectures (two in every prefecture under the channel plan)
1 in the major cities of every prefecture
(Tokushima and Saga prefectures have only 1 each as yet)

As of the end of March 1977, the number of stations, including relay stations, totalled 2,362. These stations bring 90 to 95 per cent of households in their respective areas under coverage.

(c) FM broadcasting

While FM broadcasting in Japan is implemented on a nationwide scale by NHK, the commercial broadcasters have not yet entered the stage of full-scale implementation. This arises from the policy of the Ministry of Posts and Telecommunications which envisages a future replacement of medium wave broadcasting with FM broadcasting.

Nevertheless, because the people's desire for commencement of commercial FM broadcasting has proved to be very strong, in 1968, the Ministry assigned frequencies for Tokyo, Nagoya, Osaka and Fukuoka only and licensed FM stations successively in these four areas.

These stations were licensed on condition that they undertake broadcasting that will take advantage of the character of FM broadcasting, and each station is obliged to present 70 per cent or more of its programmes in stereo.

(d) Shortwave broadcasting

In principle, Japan's commercial broadcasting follows the rule of independent operation in each region. There is one exception, however: the Nippon Shortwave Broadcasting Company, which operates shortwave broadcasting on a nationwide scale.

The Nippon Shortwave Broadcasting Company commenced its service in 1954, and, operating a transmitter in Tokyo and Sapporo, it covers all of Japan plus ships out at sea.

The broadcast programmes are of a general nature with special emphasis placed on market news and educational programmes.

4 NETWORKS

The Broadcast Law Article 52-3 provides that 'private broadcast enterprisers must not conclude any arrangement relating to supply of broadcast programmes which includes any terms by which broadcast programmes are supplied by a particular person only'.

While concentrated affiliation of commercial broadcasters is held in check through such prohibition, broadcasting of the same programme simultaneously by various commercial broadcasters is done by relaying the programme to the broadcasting stations involved from the programme producing station. However, as the individual commercial broadcaster reserves the editing right even in this case, he considers himself as not being under the control of another commercial broadcaster.

The need for networks arises from the following points: 1 It is necessary for the commercial broadcasters in the various parts of the country to maintain an agreement on nationwide coverage work, since they operate independently in their own respective region. 2 Prohibitive high costs of programme production (television programmes in particular) makes it untenable for a single company to bear it all. 3 Programme production at local stations is extremely difficult as talents and programme materials are concentrated in central cities like Tokyo and Osaka. 4 As many of the advertising sponsors want nationwide advertisements, a business structure capable of meeting their demands is required. Such a nationwide structure also links up with a lower unit price of advertisements.

The Tokyo stations serve as key stations of the networks. In radio, the Tokyo Broadcasting Systems (TBS) and stations affiliated with it form one network, while the Cultural Broadcasting Company and Nippon Broadcasting Company and their affiliates form another.

In television, four networks have been formed, including those formed by TBS and affiliates, NTV and affiliates, Fuji TV and affiliates and TV Asahi and affiliates.

Network formation in radio has not made as much

progress as in television, due to the fact that the cost of radio programme production is not prohibitive and that individual radio companies have provided themselves with programme production facilities. Programmes relayed over the networks are limited largely to such items as news and sports relays which require simultaneous and quick reporting.

Most of the television programmes, because of their nature, are networked. However, no station obtains all its programmes from a single key station. Most of the local stations obtain cross network services from a number of networks. Individual stations also present programmes they themselves produce, which consist largely of local items.

The Tokyo Channel 12 has fallen behind in network formation as it was a late-comer which started out as a station specializing in science and technology education. The station, nevertheless, is building up network relations by supplying local stations with programmes little by little.

When the commercial broadcasters obtaining 40 per cent or more of their programmes from specific key stations are grouped, 23 are found to be under TBS, 22 under Fuji TV, 17 under NTV and 7 under TV Asahi. Other stations either obtain less than 40 per cent of programmes from one key station working under the cross network system, or present mostly programmes of their own production.

The ratio of network programmes is not necessarily fixed. Rather, it is quite usual to see slight changes according to the programming policies of local stations.

When NTV and TBS pioneered Japan's commercial television service, the local stations were divided between the two companies to form networks. Subsequently, networks were formed by NET (now TV Asahi) and Fuji TV, the two key stations of these companies affiliating themselves with local stations in areas already having two or more broadcasters. For this reason, many of the stations under these two late-comers consisted of new UHF stations, and their networks suffered in comparison with those of NTV and TBS. However, as the UHF broadcasters have reached about the same level of development as the pioneer VHF broadcasters, the four networks have gradually become balanced.

Also, along with the expansion of the areas of the four commercial broadcasters, the numbers of the key and local stations have become equal in many areas, meaning that the trend towards concentrated affiliation of commercial broadcasters is gradually intensifying.

5 EXISTING STATE OF BUSINESS OPERATIONS

(a) Business revenue and expenditure

The total revenue of 107 commercial broadcasting companies in the 1975 fiscal year amounted to ¥562,900 million. Of this total, the 53 radio companies accounted for ¥71,100 million, while the 90 television companies accounted for ¥478,300 million. (As 36 companies operate radio and television concurrently, the sum of radio and television companies does not coincide with the number of commercial broadcasting companies.) The revenue from television is 6.7 times that from radio and accounts for 85 per cent of the total revenue of commercial broadcasters.

Business revenues of various commercial broadcasting companies show considerable differences which depend on such factors as the scale of the service area, economic gaps among the regions and the forms of income (ratio between income from time sales and spot sales and income from programme production). In sales levels for the fiscal year 1975, 43 companies - nearly half the total number of commercial broadcasters - recorded between ¥1,000 million and ¥3,000 million, while 10 companies achieved sales exceeding ¥10,000 million.

Table 9 gives the names and amounts of the ten companies that recorded the highest sales in the television and radio divisions. As is evident the leading earners are companies based in Tokyo and other major cities.

Income composition shows considerable differences according to region. In Tokyo, income from television programme production accounts for about 40 per cent; in Osaka and Nagoya, spot commercial sales account for 50 per cent; and at local stations, sales of time account for more than 50 per cent. High income from programme production at the Tokyo companies is attributable to the fact that, as the key stations, they serve as programme production centres.

In the case of radio, regional differences in revenue

TABLE 9 Best ten in radio and television revenue (1975 fiscal year) (unit: ¥ million)

	Television			Radio		
Rank	Company	Revenue	Share %	Company	Revenue	Share %
1	Nippon TV Network Corp.	52,089	10.9	Nippon Broadcasting Co.	8,699	12.1
2	Fuji Telecasting Co.	48,450	10.2	Tokyo Broadcasting System Inc.	7,871	11.1
3	Tokyo Broadcasting System Inc.	46,411	9.7	Cultural Broadcasting Co.	7,090	10.0
4	Asahi National Broadcasting Co.	45,197	9.5	Tokyo FM Broadcasting Co.	2,892	4.1
5	Mainichi Broadcasting System Inc.	16,660	3.5	Tokai Broadcasting Co.	2,857	4.0
6	Asahi Broadcasting Corp.	15,152	3.2	Asahi Broadcasting Corp.	2,693	3.8
7	Yomiuri Telecasting Corp.	14,233	3.0	Chubu Nippon Broadcasting Co.	2,326	3.3
8	Kansai Telecasting Corp.	13,885	2.9	Mainichi Broadcasting System Inc.	2,291	3.2
9	Tokyo Channel 12 TV	10,948	2.3	Osaka Broadcasting Corp.	1,861	2.6
10	Chubu Nippon Broadcasting Co.	9,227	1.9	Radio Kanto Co.	1,747	2.5
TOTAL	TV companies (90)	478,316		Radio companies (53)	71,126	

(Note: The above revenue excludes other business revenue)

composition are not as conspicuous as in television since revenues are dependent largely on sales of programme time and spot commercials. This is because of the low cost of radio programme production and the use of large numbers of original programmes produced at the respective company.

Profits of 107 commercial broadcasters before tax in the fiscal year 1975 totalled ¥45,900 million, representing a decline of 2.5 per cent from the ¥47,400 million registered by 105 companies in the preceding year.

While commercial broadcasting has gone through slight ups and downs since its inauguration on account of changes in the economic environment, it has generally continued to enjoy smooth progress. From the latter half of 1973, however, because of the impact of the oil crisis, a dulling of the growth rate of revenues and at the same time, an increase in running expenses, resulted in reduced profits for four consecutive business terms. A slight stabilization of economic conditions in 1976 has given rise to expectations for increased revenues and profits, but the view that the future of commercial broadcasting cannot perhaps be looked upon optimistically is gaining momentum, since a high level of economic growth such as that which sustained the initial development of commercial broadcasting, can no longer be expected for Japan.

Recent distribution of profits by scale is given in Table 10. The largest number of companies (48) gained profits ranging from ¥100 million to ¥500 million, but as many as 16 companies, in particular, late-comer UHF companies and radio-only companies, show losses.

Changes in broadcast revenue and income for commercial broadcasters over the last five years are as shown in Appendix 3.2.

(b) Capital and assets

The total capital of 106 commercial broadcasters (Kyokuto Hoso excluded) amounted to ¥55,100 million as of the end of 1975. The distribution of capital by scale is as shown in Table 11. 74 companies with capital less than ¥500 million account for the majority. Naturally, the capital of the later-comer UHF television stations are relatively larger than those of companies established earlier.

Private capital (including shares, various reserve

TABLE 10 Distribution of profits by scale (fiscal year 1975)

Profit before tax (year)	Radio-TV co.	Radio-only co.		TV-only co.		Total
		Medium, short-wave	FM	VHF	UHF	
Up to ¥10 million	–	–	–	–	–	–
10 ∿ 50 million	1	4	–	1	1	7
50 ∿ 100 million	3	–	1	–	–	4
100 ∿ 500 million	15	3	2	3	25	48
500 ∿ 1,000 million	12	1	1	5	4	23
1,000 ∿ 2,000 million	3	–	–	2	1	6
2,000 ∿ 3,000 million	–	–	–	1	–	1
3,000 ∿ 4,000 million	–	–	–	1	–	1
4,000 million above	1	–	–	–	–	1
TOTAL	35	8	–	13	31	91
Deficit companies	1	5	–	1	9	16

TABLE 11 Capital

Capital	Radio-TV co.	Radio-only co.		TV-only co.		Total
		Medium, short wave	FM	VHF	UHF	
Up to ¥100 million	-	2	-	-	-	2
100 ∿ 300 million	17	6	3	4	1	31
300 ∿ 500 million	9	2	-	3	26	41
500 ∿ 1,000 million	7	2	-	2	10	21
1,000 ∿ 2,000 million	2	-	-	3	3	8
2,000 million or above	1	-	-	2	-	3
TOTAL	36	12	4	14	40	106

surpluses and net profits for current term) available to the operators amounted to ¥222,100 million at the 106 companies as of the end of March 1976. This was four times the total capital, indicative of considerable reserve assets. The reserve assets, however, differ according to the company. Seen from the multiple of private capital against share capital, the highest ratio of six times is found in the radio-television operating group while the lowest ration of 1.3 times is found in the UHF television group. The total assets of the 106 commercial broadcasters amounted to ¥495,700 million as of the end of 1976, an average of ¥4,700 million per company. The largest was TBS with ¥59,100 million and the smallest Tochigi Broadcasting Company at ¥1,530 million. Eight companies have capital exceeding ¥10,000 million.

(c) Personnel

The 107 commercial broadcasting companies employed a total of 25,381 people as of the end of 1975. The distribution of employees among the various services is as given in Table 12. The UHF television stations and FM radio stations employing less than 200 workers all follow the policy of operation by small staff.

TABLE 12 Staff (as of the end of the fiscal year 1975)

Staff size		Radio-only co.		TV-only co.		Total
	Radio-TV co.	Medium short wave	FM	VHF	UHF	
Less than 100	1	6	3	-	20	30
100 ∿ 200	8	4	1	4	20	37
200 ∿ 300	12	3	-	3	-	18
300 ∿ 500	9	-	-	1	-	10
500 ∿ 1,000	5	-	-	3	-	8
1,000 or more	1	-	-	3	-	4
TOTAL	36	13	4	14	40	106

6 CHANGES AND EXISTING STATE OF ADVERTISING EXPENDITURES

The gross national product of Japan in 1951, when commercial broadcasting was inaugurated, amounted to only

¥5,400,000 million, and the total expenditure for advertising accounted for ¥24,300 million, which was no more than 0.5 per cent of the gross national product. With economic recovery, however, advertising expenditure gradually increased, and this increase in turn, resulted in stimulating individual consumption and equipment investment, serving as a factor in the subsequent high-level growth of the Japanese economy.

The ratio of total advertising costs against gross national product is as shown in Appendix 4.1. The ratio climbed to a peak of 1.22 per cent in 1963, but has gradually declined since, recording 0.85 per cent in 1975. The same year's total spending on advertisements, however, amounted to ¥1,200,000 million against a gross national product of ¥144,900,000 million. Absorbing this remarkable increase in spending for advertisements commercial broadcasting has followed a course of rapid development.

Among the mass media used for advertising, newspapers, magazines, radio and television are the leading four, accounting for 77 per cent or more of total advertising expenditures.

Prior to the inauguration of commercial broadcasting, newspapers, with their longer history, had overwhelmingly dominated the advertising world, accounting for 70 per cent of total spending on advertisements. When commercial broadcasting made its appearance however, it gradually made inroads into the newspaper advertising area, and in 1965, spending on radio and television advertisements reached 36.9 per cent of the total, higher than the 35.8 per cent spent on newspaper advertisements. In 1975, moreover, advertisements on television alone accounted for 34 per cent of the total against the newspapers' share of 33.1 per cent. Television has thus become the king of advertising media.

In 1975, spending on advertisements involving broadcasting totalled ¥480,000 million. Of this total, radio accounted for ¥60,000 million and television, for ¥420,000 million, accounting respectively for 5 per cent and 34 per cent of total spending on advertisements.

The composition of spending on radio and television advertisements by type of business enterprise is as shown in Table 14. While the transportation equipment, food, beverage, service and amusement business constitute leading advertisers on radio, businesses advertising largely on television are those in the food, beverage, cosmetics, cleaner, medicinal and home electric appliance fields. Newspaper advertisements are dominated by those for housing, building materials, publications, department stores, service business and motion pictures. The character of the media obviously makes for considerable differences in the type of advertiser.

TABLE 13 Advertising expenditure (1975)

Media	Amount (in ¥100 million)	Component ratio %
Newspapers	4,092	33.1
Magazines	670	5.4
Radio	602	4.9
Television	4,208	34.0
TOTAL for 4 media	9,572	77.3
Direct mail	525	4.2
Outdoor etc.	2,024	16.4
Export ad.	254	2.1
TOTAL excluding 4 media	2,803	22.7
Total spending	12,375	100.0

7 BROADCAST PROGRAMMES

(a) Programme compilation and production

As stated before, most of the commercial broadcasters are general broadcast programme stations which are obliged to present news, educational, cultural and entertainment programmes in a balanced form.

In radio, music programmes are predominant, accounting for 40 to 50 per cent of the total, followed by educational, cultural and news and information items. During the professional baseball season, incidentally, most radio stations schedule relays of the games in the evenings, but sports programmes through the year account for about 10 per cent.

In television, about half the total broadcasting time is taken up by entertainment programmes, followed by 30 to 40 per cent of educational and cultural programmes and 10 per cent of news and information.

TABLE 14 Advertising expenditures in the 4 media by business enterprise (1975) (unit ¥10 million)

	Media				
Type of business	Newspaper	Magazine	Radio	TV	Total
Basic materials	245	67	241	673	1,226
Foods, beverages	2,373	804	666	12,240	16,083
Medicinal chemicals	1,514	422	142	3,322	5,400
Cosmetics, cleaners	777	898	149	4,378	6,202
Clothing, personal articles	655	643	121	1,319	2,738
Publications	4,748	241	490	432	5,911
General industrial equipment	450	67	66	732	1,315
Precision and business machines	1,187	355	125	1,561	3,228
Electrical equipment	1,473	536	361	2,090	4,460
Transportation equipment	1,473	208	942	1,019	3,642
Home utensils equipment	1,269	335	175	2,395	4,174
Housing, building materials	6,731	214	149	1,398	8,492
Wholesale, department stores	3,632	174	393	1,234	5,433
Financial, insurance	2,158	402	136	570	3,266
Service, amusement	3,806	523	548	2,677	7,554
Government offices, organizations	491	150	173	1,004	1,818
Medical, educational, etc.	7,938	661	1,143	5,036	14,778
TOTAL	40,920	6,700	6,020	42,080	95,720

The primary task of the programming division is to decide what types of programmes should be presented in what form to what classes of audience, and to determine time scheduling for such programmes. The programming policy calls for the drawing up of broad frameworks for the day, the week and the season, while taking into account factors related to the audience (such as people's living habits, programme selection trends and preferences), factors related to business operations arising from the nature of commercial broadcasting, and various other factors such as trends at other stations and so forth.

Once the programming policy is adopted, the production division undertakes production of individual programmes in accordance with the policy. Variation is provided in the process of programme production depending on differences in programme items, such as social and information programmes including news, news commentaries, forums and documentaries, and educational and cultural programmes like lecture courses and comments. In all cases, however, the basic processes of planning, production and transmission are the same.

(b) Daily broadcasting hours

The average daily broadcasting hours of radio stations in Japan are about 22 hours and 30 minutes. Midnight radio broadcasting was inaugurated by Tokyo stations in 1959, and this service has now expanded to local stations. By 1975, there were 20 stations out of the total of 53 which were broadcasting for more than 23 hours a day.

Daily television broadcasting hours are longer at stations in major cities like Tokyo and Osaka; the average for the whole of Japan is about 17 hours. Most stations in larger cities operate midnight broadcasting of films and the like. At these stations, daily hours of broadcasting reach about 22 hours. Extension of broadcasting hours has not been seen in the past several years for a number of reasons, including voluntary suspension of midnight broadcasting for a time as a result of the electric power saving policy adopted in the wake of the 1973 oil crisis.

(c) Broadcast advertising

Commercial messages (CMs) are inserted into commercial broadcast programmes. While CMs have aspects that are helpful in everyday life and some enjoy popularity among audiences for their originality, on the whole, CM insertions tend to become too frequent since there is no special regulation regarding their quantity or insertion time zones. And this, in turn, is likely to lead to interference with the broadcasting programmes. For this reason, the Broadcasting Standards of NAB (see later) provide for voluntary discipline such as limiting the CM time to within 18 per cent of the total broadcasting time.

CMs are normally classified as follows:

1 Programme CMs inserted by the advertiser during presentation of the programme he sponsors. These are arranged to be aired before the programme begins, during programme presentation and at the end of the programme.

2 Spot CMs presented between programmes (station break) with no connection to the actual programmes. These range from 5 seconds to 60 seconds in duration.

From the audience viewpoint, CMs sometimes seem very long since often a number of spot CMs will follow the CMs at the end of one programme and then other CMs will come on before the beginning of the next programme.

3 Participating CMs. Due to rising costs of programme production, the number of advertisers singly sponsoring programmes has decreased and cases where two or more sponsors share or participate in a programme advertisement have increased. In this case, a plural number of CMs are inserted during the programme presentation in a spot CM style. A CM of this type is known as the PT-CM.

4 Classified CMs. Time frames other than the station break time are scheduled for exclusive use of spot CMs. CMs presented in this time frame are known as classified CMs.

In general, the production of a CM is entrusted by the advertiser to an advertising agent, who undertakes planning and production of the CM and sends it to the broadcasting station concerned. Most CMs are produced in film and the sound is optically recorded. The use of film is because the broadcasting station has to print CMs in accordance with the frequency of insertion and films are easier to print. Lately, however, VTR-CMs, which are even easier to produce, are beginning to be used.

Also, while most CMs are presented in animation, local sponsors still use large numbers of still pictures presented as superimposed images.

(d) Colour programmes

Presentation of programmes in colour at the four stations in Tokyo began in 1960. Thereafter, the microwave relay channels greatly improved in standard throughout the country and by 1966, stations broadcasting colour programmes were operating throughout the country. By 1968, all the commercial stations were presenting their programmes in colour.

8 MEASURES TO IMPROVE BROADCAST PROGRAMMES

The Broadcast Law obliges the commercial broadcasters to establish broadcast programme consultative committees. These are entrusted with the task of deliberating on the programme standards and basic programming policies and presents its opinion when it sees the need for improving broadcast programmes. In such a case, the broadcaster must respect its views and take necessary measures.

This provision was added to the Broadcast Law in 1959 when, along with the wider diffusion of television, the vulgarization of programmes and their demoralizing influence on the minds of young people became an issue.

Each commercial broadcaster has set up a programme consultative committee in accordance with the provision and is mindful of self-discipline.

Meanwhile, the Broadcasting Standards of the National Association of Commercial Broadcasters in Japan (NAB) have been established as guidelines in order to assure effective self-discipline. Also, for the purpose of assuring observance of these standards, the Broadcasting Standards Council has been set up as a body directly responsible to the NAB president.

The following bodies are also maintained to promote further improvement of broadcast programmes:

Broadcaster's Council for Better Programming

This was established for the purpose of further promoting voluntary efforts towards programme improvement in co-operation between NHK and NAB. The council is composed of seven members who are well-informed and cultured people having no connection with the broadcasting organizations. The purpose of the council is to study public opinion related to broadcast programmes, discuss means of programme improvement, co-operate with all the broadcasters in their voluntary efforts towards programme improvement and the strengthening of broadcast ethics and the development of broadcasting culture as a whole.

Broadcast Programming Centre of Japan

This was set up to facilitate the compilation of educational programmes by commercial broadcasters. It is maintained jointly by NHK, commercial broadcasters, advertising agents and electrical equipment makers.

Its principal functions are, 1, planning and production, 2, maintenance of the programme library and, 3, programme exchange with overseas organizations in connection with educational and cultural items.

Japan Advertising Review Organization

This was established jointly by the commercial broadcasters and advertisers to process complaints related to advertisements and exhibitions.

All Japan Radio and Television Commercial Council

This organization is maintained jointly by the commercial broadcasters and advertisement sponsors for the purpose of uplevelling voluntary regulations of CMs.

9 AUTOMATION IN BROADCASTING

The commercial broadcasters are active in promoting the introduction of automatic systems for operating unattended transmitters and relay stations, for speeding up programme compilation, for assuring accurate programme transmission and for streamlining business operations. While the extent of automation differs according to the company, automation systems are now being adopted in the following sectors:

(a) Remote control of broadcasting equipment

In both radio and television, the studios are generally found in the same building as the head office, but it is customary to operate the transmitter away from the offices in places where transmitting conditions are better. The relay stations are, in many cases, set up in remote areas and hilly districts. In the case of television, in particular, many such stations are built on hilltops and access is difficult. Research work on the remote control of transmitting equipment, improvement of the reliability of transmitting equipment and the automatic monitoring thereof, and on the installation of spare equipment, therefore, has been undertaken since early days. And, with the achievement of adequate technological developments, most such stations are now unattended facilities, controlled remotely from the head offices.

(b) Automation in programme transmission

Automation in programme transmission has been desired since early days in order to assure speedy and accurate insertion of CMs during programme presentation and station breaks. As a means to this end, peripheral equipment such as the automatic opaque card projector feeding device, and, in the case of film, the automatic film projector feeding device and the automatic vidicon camera adjusting device have been developed.

From around 1965, moreover, the coupling of these devices with computer systems has made possible an automatic interswitching between such devices and studios and VTRs, thus broadly automating the programme transmission mechanism.

While the taking-off point for computerization in commercial broadcasting was programme transmission, large-scale computers subsequently introduced are now used for in-service data processing, ranging from business inputs to programme transmission, thus computerizing a series of operations including sales, programming and finance.

(c) Automatic switching of network programmes

For stations operating under the cross network system, interswitching of network programmes is necessary. Also, as all the stations present programmes of their own production in addition to network programmes, interswitching between network programmes and original programmes is necessary.

In order to ensure accurate switching and simplification of operation, cue signals are attached to network programmes at the key station so as to facilitate automatic switching between programmes.

10 NATIONAL ASSOCIATION OF COMMERCIAL BROADCASTERS IN JAPAN (NAB)

This was established as a voluntary organization on 20 July 1951 by the sixteen commercial broadcasters then in operation. On 21 April 1952 it was organized into a foundation under authorization of the Radio Regulatory Commission, the competent government agency.

The principal objects of NAB are the strengthening of broadcast ethics, the furtherance of public welfare through broadcasting and the progress and development thereof, and the processing of problems common to commercial broadcasters. All the commercial broadcasters in Japan are members of NAB. NAB undertakes the following businesses:

1 Establishment and strengthening of broadcasting ethics;
2 Contact among members and managing of problems common to all;
3 Research and investigation related to broadcast programmes, engineering and management;
4 Contact with the Diet, government offices and others in connection with problems involving broadcasting;
5 Preservation of official archives as prescribed by the Copyright Law;
6 Education and public relations connected with broadcasting;
7 Publication of industry paper;
8 Functions necessary for promoting welfare, goodwill and harmony of commercial broadcasters;
9 Other operations deemed necessary to accomplish the aims of NAB.

NAB's businesses are administered by the Board of Directors which meets once a month. For professional inquiries into related problems the Broadcasting Standards Council is set up under the President. Various committees (general affairs, legislation related to broadcasting, sales, radio, copyright, engineering, labour, finance, budget, commercial broadcasters' conference, programme) are set up under the Board of Directors.

11 TRAINING OF STAFF

Staff training is undertaken to cope with the everchanging world of broadcasting. Training courses include those offered by NAB and those given by individual broadcasters.

Training by individual broadcasting companies consists mainly of sending trainees to overseas countries and participation in seminars conducted by outside organizations.

12 INTERCHANGE WITH OVERSEAS COUNTRIES

Interchange between commercial broadcasters and overseas countries is increasing yearly. In addition to station level activities, such as agreements for exchanging programmes and news, and affiliations such as sister stations, activities on a level involving all stations, including NHK (for instance, the 'Japan-US Television Programme Festival'), are achieving important results in the international exchange of broadcasting culture.

The 'Japan-US Television Programme Festival' has been held alternately in Japan and the USA every year since 1973. The 'programme trade fair' for purchase and sale of programmes and visits to sister stations constitute its principal activities, but the fourth festival held in Japan in January 1977 included in addition to programmes, working committees on advertising, research and engineering, thus expanding the sphere of activity.

Sister stations relations with American stations are now maintained by fifteen Japanese commercial broadcasters, and the number is likely to increase. The longest-standing relations were established in 1964 between RKB Mainichi Broadcasting Company of Japan and KTVU of Oakland, California. Since then this number has rapidly increased, with the Japan-US TV Programme Festival as the

TABLE 15 Principal Training Courses

Classification	No. of times per year	No. of days	No. of participants	Remarks
Training in basic TV engineering	1	Approx. 10	Approx. 60	Entrusted to NHK
Training in sound engineering centred on mixing	1	" 6	" 60	Training Institute
Training in TV programme production techniques	1	" 4	" 60	
Sales seminars for coping with new industrial and economic environments	1	" 2	" 200	
Radio seminars for developing media value of ratio	1	" 2	" 100	
Training in processing problems related to copyright	1	" 2	" 50	
Training in labour problems	Several times	2 per session	Approx. 100 per session	

momentum. Aside from mutual co-operation between the sister stations, such as interchange of programmes and personnel, their relations are also achieving good results in the promotion of goodwill between the cities in which the stations are located.

Interchange with overseas organizations through conclusion of 'business agreements' and 'co-operative agreements' is increasing notably. For instance, TBS, a television company based in Tokyo, has a news exchange agreement with CBS of the USA and has exchanged protocol with the State Committee for Television and Radio Broadcasting of the Council of Ministers of the USSR; NTV, another television company in Tokyo, maintains news exchange agreements with NBC of the USA, CBC of Canada and the Toyo Broadcasting Company of Korea; TV Asahi, also of Tokyo, has business and news agreements with ABC of the USA and has exchanged protocols with the State Committee of the USSR; Fuji TV, also of Tokyo, has co-operative agreements with broadcasting organizations in seven countries, including the State Committee of the USSR and Taiwan TV. In addition to programme exchanges and news agreements, examples of joint production of programmes are also increasing.

Chapter 6

THE FUTURE OF BROADCASTING

1 MAINTENANCE AND DEVELOPMENT OF THE EXISTING SYSTEM

Broadcasting in Japan has been enriched, both in quality and quantity, to a degree not often seen in the world. Because the present broadcasting system has shown such good results in the past, it is considered worthy of continuation into the future. As ground broadcasting networks go, the existing system has reached the stage of the completion of the finishing touches, such as elimination of the few remaining areas not accessible to television service and the adequate provision of FM broadcasting.

Meanwhile, how new broadcasting media, such as satellite broadcasting, multiplex broadcasting and still-picture broadcasting, which have been made available as a result of technological innovations, can be incorporated into the existing system in order to meet the people's new demands for information and make broadcasting more fruitful, has become the most important question that faces Japan today.

2 SATELLITE BROADCASTING

Activities related to satellite broadcasting have advanced on the basis of the decision to launch an experimental satellite in March 1978 and the international agreement, at the 1977 World Administrative Radio Conference, on plans for the use of the 12GHz band for operational satellite broadcast systems. As a result, practical application tion of satellite broadcasting has reached the stage of appearing on concrete schedules.

As of now, NHK's television programmes reach 97.7 per

cent of the total population, while commercial television programmes reach 94 per cent. However, as the desire of the remaining few per cent of inhabitants for access to television is strong, the government considers it highly important to take measures that will lead to availability of television in every corner of the country. More stations are being established (7,300 television stations have already been built, for instance), but the remaining areas where reception is difficult are becoming increasingly smaller and more widely scattered. As a result, both NHK and commercial broadcasters are experiencing markedly rising costs in their attempts to eliminate reception difficulties. In the cities, too, the conglomeration of high-rise buildings is creating obstacles to reception through reflection or obstruction of radiowaves. In particular, the advent of skyscrapers has increased the difficulties, and amelioration of the situation has become difficult, even with the existing cable television systems.

As a consequence, the solution now being considered is through the direct reception of satellite broadcasting. As a frequency as high as 12 GHz is used in satellite broadcasting, not only is it possible to overcome ghost interference but also selection of appropriate orbit position allows an adequate angle of elevation. Since satellite broadcasting is a thus superb means of avoiding reception difficulties its development to the stage of practical application is being strongly urged.

Also, with advances and diversification in social life, resulting from rapid development of the social economy, the need for making use of television as a means of re-educating the general public and providing educational opportunities for young working people has become ever more acute. Satellite broadcasting would make possible the provision of programmes for less than the cost of the ground system and would allow allocation of new channels for educational broadcasts because of the use of new frequencies.

In order to meet the people's strong demand for satellite broadcasting, Japan is scheduled to launch a medium-scale broadcast satellite in March 1978. This satellite is to be used for experimenting with operating techniques and sound and video transmissions needed. Future plans, if realized, would make possible individual reception from a large-scale broadcast satellite.

Meanwhile, with the development of the satellite system highly sensitive and low-cost receivers (S-U converters) are considered necessary. As a result of intensive research work at NHK and elsewhere, an S-U converter composed of a mounted planar circuit in waveguide, which is simple and can be cheaply mass-produced, has been developed.

The prospects of solving the various technical problems have therefore become brighter, but the application of satellite broadcasting is inevitably bound to give rise to international problems, since it makes use of space and its coverage area is very wide. Foreseeing such problems, for five weeks from 10 January 1977, a World Administrative Radio Conference for the planning of the broadcasting-satellite service in the 12GHz Band (WARC-BS) was held in Geneva, and efforts were made towards drafting a technical plan for the use of frequencies and orbital positions. Efforts towards realizing satellite broadcasting are also to be made in conjunction with international conferences, such as the Legal Sub-Committee of the Committee on the Peaceful Uses of Outer Space, which continue to deliberate on the governing principles of the broadcasting satellite law.

3 UNIVERSITY ON THE AIR

In Japan, 93 per cent of those who complete the compulsory nine years of education advance to high schools and 37 per cent of high school graduates advance to colleges and universities. These rates are quite high even by world standards. However, the growing complexity and advances in social life have led to the demand for opportunities for higher education from young working people and from those already in the business world.

While Japan's broadcasting system is carefully planned, broadcasting centred on educational programmes is implemented only by one television network and one radio network, both of NHK. These stations, moreover, are concerned largely with secondary and social education, and present university courses for only one hour a day.

Because of this the Ministry of Posts and Telecommunications and the Ministry of Education jointly adopted a policy in 1969 for promoting research and investigation into the establishment of a university on the air which, of new method and form, would make use of the television and FM broadcasting information media which impose neither time, geographical nor age limits to the undertaking of university studies. Since then, the two ministries have completed their investigations into the educational research system, curriculum, educational methods, form of establishment and construction costs of such a university. This plan has now reached the experimental stage using television and radio stations of commercial broadcasters.

The university on the air calls for coverage of 90 per cent of the total population with UHF television and FM radio networks, but as construction of the two networks is estimated to cost some ¥42,000 million, nationwide broadcasting of university programmes is likely to be put off till some future time.

4 MULTIPLEX BROADCASTING

The demand for private wireless stations has greatly increased recently. The approximately 5,000 wireless stations in operation as of 1950 had jumped to 1.3 million by the end of March 1976.

The lack of available frequencies constantly plague those concerned. This has led to an intense interest in the development of more efficient methods of radiowave utilization or the development of new frequencies.

The multiplex broadcasting system is a means of broadcasting by which new signals are superimposed over the gaps in the radiowaves used for television or FM broadcasting. This can, indeed, be considered an effective and efficient way of utilizing radiowaves. Moreover in the implementation of multiplex broadcasting, most of the equipment needed on the transmitting side can be filled with existing broadcasting equipment. On the receiving side, too, attachment of simple adapters to the existing receiving sets allows reception, thus making it very economical (facsimile broadcasting is excluded, however).

Study of experimental multiplex broadcasting has been undertaken in Japan since 1958. It has become possible to superimpose sound multiplex broadcasting, written information broadcasting, still-picture broadcasting and facsimile broadcasting on television and FM broadcasting. There are prospects for multiplexing several kinds of other programmes on a single television channel and one kind on the single FM channel but further study related to compatibility with existing broadcasting and to standardization of the system is being promoted.

In 1974, the Ministry of Posts and Telecommunications set up the Research and Investigation Conference Related to Multiplex Broadcasting, looking into the institutional aspects, such as what legislation would be needed for implementing multiplex broadcasting, and as a result of studies undertaken by knowledgeable members of the conference, a report has been submitted. The biggest issue in the report proved to be whether or not some third party other than the existing broadcasters should be allowed to operate independent programmes multiplexed on existing television and FM broadcasting. Authorization of the use of multiplex broadcasting for independent programmes would meet the people's demand for diversification and improvement of and increased volume of information, thus achieving maximum effect, but concern was expressed regarding overconcentration of the information media should the management of multiplex broadcasting become an additional undertaking of existing broadcasters. In this connection, the report states that, as a general direction, the concentration of media should be avoided. Concrete measures regarding the matter, therefore, appear to call for further study.

5 CATV

CATV is expected to become more important, as a multi-purpose information medium. In this connection, recently, there have been a number of noteworthy developments.

Previously, the Coaxial Cable Information System (CCIS) Study Committee had been set up in the Ministry of Posts and Telecommunications and the future potential of CATV as CCIS was studied. Following this study, an actual CCIS was built in a living community for all sorts of purposes. Then, from January 1976, experiments on CCIS were started in Tama New Town in the suburbs of Tokyo, carried out by Living-Visual Information System Development Association, to probe into the needs of inhabitants and undertake technical experiments in connection with multi-purpose utilization of CATV. The results of this experiment are worthy of close attention in what they reveal of the possibilities of CATV as a multi-purpose information medium. The Foundation also has plans for conducting an experiment from 1978 at Higashi Ikoma in the suburbs of

Osaka. In this experiment, optical cables are to be used as the transmitting route and a two-way information system connected with a computer system is to be tried out.

Meanwhile, as part of agriculture and forestry administration, the development of a pluralistic CATV information system for farming areas is being planned to promote modernization of agricultural production and improvement of the living environment of rural societies by making available systematic and comprehensive information on production, distribution and consumption. For the time bein being, only retransmission of television programmes and broadcasting of original programmes is planned, but in the future, automatic control of farming facilities and all kinds of two-way services are to be undertaken.

Over recent years, various newly developed towns have been introducing the CATV system as incidental equipment for large-scale housing complexes, and, in addition to community reception of television programmes, the broadcasting of independent programmes, anti-disaster and anti-crime services and automatic meter reading services for electric power, gas and water supplies are being planned. Since October 1976 unique experiments have been being made in the study of various problems attending the introduction into rural areas of a video transmission system, using the paired cables laid for cable radio broadcasting or cable broadcasting telephone service. Although only one channel is available, the future of this system should be watched closely as a possibility leading to the development of a CATV system which would provide indepeddent, low cost broadcasting, utilizing already available equipment.

APPENDICES

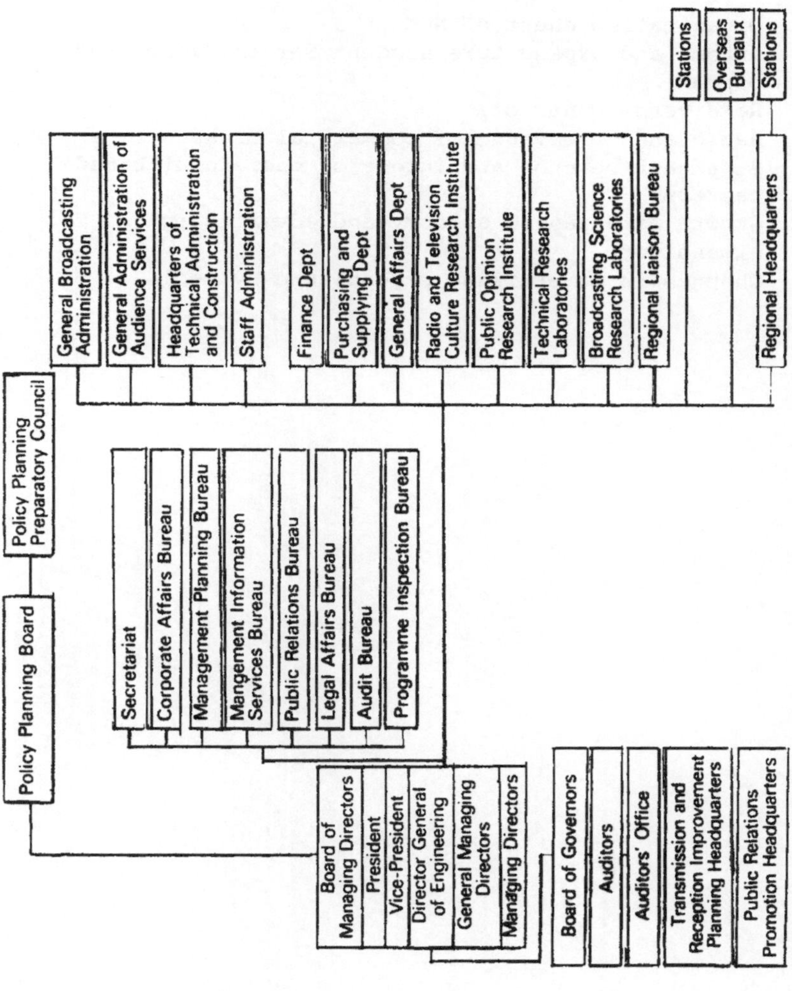

APPENDIX 1 Organization chart of NHK (as of August 1977)

APPENDIX 2.1 Income and expenditure account for the years 1972-7 (in ¥ million)

	1972-3	1973-4	1974-5	1975-6	1976-7
Business income	109,979	118,723	125,786	131,374	191,505
Receivers' fees	107,928	116,010	122,475	128,371	187,523
Grants	231	203	293	352	612
Others	1,820	2,510	3,018	2,651	3,370
Business expenditure	110,545	119,679	129,834	149,344	170,215
Salaries	36,902	41,668	49,776	58,207	63,772
Domestic broadcasting cost	29,964	30,100	31,074	35,833	42,304
Overseas broadcasting cost	763	791	811	865	1,043
Audience service cost	11,779	13,315	14,732	18,774	23,378
Research and investigation cost	1,574	1,875	1,711	1,803	2,090
Administrative cost	11,637	12,563	14,305	17,238	20,765
Depreciation cost	15,469	16,332	13,934	12,965	12,955
Financial cost	2,457	3,035	3,491	3,659	3,908
Balance	- 566	- 956	- 4,048	- 17,970	21,290
Extraordinary income (Gains from sale of fixed assets, etc)	329	31,548	1,036	593	755
Extraordinary expenditure (Loss from sale of fixed assets, etc.)	397	12,724	1,011	1,529	1,532
Total balance	- 634	17,868	- 4,023	- 18,906	20,513

APPENDIX 2.2 Receivers' contracts (the end of fiscal years 1972-6)

	1972	1973	1974	1975	1976
Total	24,433,463	24,924,985	25,753,396	26,544,758	27,058,881
Ordinary contracts	15,630,946	18,335,615	20,543,694	22,262,448	23,309,448
Colour contracts	8,802,517	6,589,370	5,209,702	4,282,310	3,749,433

APPENDIX 3.1 Radio and TV revenue of commercial broadcasters (in ¥ million)

Division	Radio revenue					TV revenue				
	1972-3	1973-4	1974-5	1975-6	1976-7	1972-3	1973-4	1974-5	1975-6	1976-7
Radio-TV co.	27,244	32,509	36,417	38,895	45,227	138,007	161,912	178,633	190,837	224,601
Radio-only co.	22,682	28,091	30,220	32,231	36,343					
TV-only co.						192,672	239,845	262,634	287,479	350,405
TOTAL	49,926	60,600	66,637	71,126	81,570	330,679	401,757	441,267	478,316	575,006

(Note: The above revenue excludes other business revenue. Four Radio-TV companies' revenue in the figures of 1976-7 is an estimate.)

APPENDIX 3.2 Broadcast revenue and income of commercial broadcasters (in ¥ million)

Division	Broadcast revenue					Broadcast income				
	1972-3	1973-4	1974-5	1975-6	1976-7	1972-3	1973-4	1974-5	1975-6	1976-7
Radio-TV co.	168,841	199,465	218,911	235,134	274,909	23,476	28,249	21,334	20,134	33,869
Radio-only co.	23,452	28,811	31,268	34,295	40,166	2,571	3,557	2,134	1,947	3,659
TV-only co.	195,130	242,841	266,861	293,509	356,203	21,083	32,058	25,549	26,563	49,787
TOTAL	387,423	471,117	517,040	562,938	671,278	47,130	63,864	49,017	48,644	87,315

(Note: Broadcast Revenue includes other business revenue. Four Radio-TV companies' revenue and income in the figures of 1976-7 are estimates.)

APPENDIX 4.1 Growth of Japanese economy and advertising expenditure (1951-76)

Year	GNP (in Y ¥ million)	Ad. expenditure (in ¥100 million)	Ratio of Ad. expenditure to GNP %
1951	54,368	243	0.45
1952	62,368	385	0.62
1953	75,264	491	0.65
1954	78,246	550	0.70
1955	88,646	609	0.69
1956	99,509	745	0.75
1957	112,489	940	0.84
1958	117,850	1,065	0.90
1959	136,089	1,456	1.07
1960	154,992	1,740	1.12
1961	191,255	2,110	1.10
1962	211,992	2,435	1.15
1963	244,640	2,982	1.22
1964	289,317	3,491	1.21
1965	319,564	3,440	1.08
1966	368,222	3,831	1.04
1967	435,657	4,594	1.05
1968	515,935	5,321	1.03
1969	596,894	6,328	1.06
1970	707,309	7,560	1.07
1971	792,536	7,868	0.99
1972	906,117	8,782	0.97
1973	1,110,610	10,768	0.97
1974	1,324,728	11,695	0.88
1975	1,449,150	12,375	0.85
1976	1,645,400	14,568	0.89

(Source: Dentsu Ad. Agency)

APPENDIX 4.2 Changes in advertising expenditure by media (1950-76) (in ¥100 million)

Media	Newspaper		Magazine		Radio	
Year	Ad. expen- diture	Ratio to previous year %	Ad. expen- diture	Ratio to previous year %	Ad. expen- diture	Ratio to previous year %
1950	120	150.6	7	140.0		
1951	180	150.0	10	142.9	3	-
1952	270	150.0	18	180.0	22	733.3
1953	320	118.5	25	138.9	45	204.5
1954	322	100.6	30	120.0	74	164.4
1955	337	104.7	35	116.7	98	132.4
1956	405	120.2	40	114.3	130	132.7
1957	510	125.9	50	125.0	150	115.4
1958	525	102.9	55	110.0	157	104.7
1959	618	117.7	80	145.5	162	103.2
1960	684	110.7	100	125.0	178	109.9
1961	824	120.5	125	125.0	178	100.0
1962	922	111.9	144	115.2	173	97.2
1963	1,120	121.5	169	117.4	171	98.8
1964	1,297	115.8	195	115.4	170	99.4
1965	1,233	95.1	192	98.5	161	94.7
1966	1,337	108.4	211	109.9	169	105.0
1967	1,611	120.5	255	120.9	195	115.4
1968	1,884	116.9	298	116.9	233	119.5
1969	2,250	119.4	348	116.8	291	124.9
1970	2,653	117.9	418	120.1	345	118.6
1971	2,681	101.1	445	106.5	388	112.5
1972	3,024	112.8	478	107.4	428	110.3
1973	3,721	123.0	572	119.7	496	115.9
1974	3,945	106.0	626	109.4	554	111.7
1975	4,092	103.7	670	107.0	602	108.7
1976	4,550	111.2	797	119.0	704	116.9

APPENDIX 4.2 (continued)

Media	Television		D	M	Outdoor etc.	
Year	Ad. expen-diture	Ratio to previous year %	Ad. expen-diture	Ratio to previous year %	Ad. expen diture	Ratio to previous year %
1950					40.5	202.5
1951					50	123.5
1952					75	150.0
1953	1	-			100	133.3
1954	4	400.0			120	120.0
1955	9	225.0			130	108.3
1956	20	222.2			150	115.4
1957	60	300.0			170	113.3
1958	105	175.0			210	123.5
1959	238	226.7			334	159.0
1960	388	163.0			372	111.4
1961	539	138.9	90	-	321	-
1962	690	128.0	102	113.3	353	110.0
1963	899	130.0	117	114.7	441	124.9
1964	1,081	120.2	115	98.3	546	123.8
1965	1,110	102.7	116	100.9	539	98.7
1966	1,247	112.3	168	144.8	602	111.7
1967	1,509	121.0	204	121.4	700	116.3
1968	1,745	115.6	213	104.4	799	114.1
1969	2,042	117.0	229	107.5	988	123.7
1970	2,445	119.7	268	117.0	1,213	122.8
1971	2,594	106.1	297	110.8	1,238	102.1
1972	2,841	109.5	388	130.6	1,417	114.5
1973	3,522	124.0	453	116.8	1,756	123.9
1974	3,917	111.2	493	108.8	1,896	108.0
1975	4,208	107.4	525	106.5	2,024	106.8
1976	5,093	121.0	818	155.8	2,296	113.4

APPENDIX 4.2 (continued)

Media	Export Ad.	
Year	Ad. expen-diture	Ratio to previous year %
1950		
1951		
1952		
1953		
1954		
1955		
1956		
1957		
1958	13	-
1959	24	184.6
1960	18	75.0
1961	33	183.3
1962	51	154.5
1963	65	127.5
1964	87	133.8
1965	89	102.3
1966	97	109.0
1967	120	123.7
1968	149	124.2
1969	180	120.8
1970	218	121.1
1971	225	103.2
1972	206	91.6
1973	248	120.4
1974	264	106.5
1975	254	96.2
1976	310	122.0

(Source: Dentsu Ad. Agency)
(Note: In 1950-60, ad. expenditure of Outdoor etc., includes that of DM.)